M**THE**O**LD**

THE THING THAT CHALLENGED
THE SUPREMACY OF MANKIND

BY GARY GENTILE

I0642574

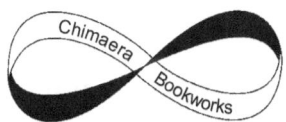

Chimaera Bookworks

Chimaera Bookworks
P.O. Box 57137
Philadelphia, PA 19111

Additional copies of this book may be purchased from the same address by sending a check or money order in the amount of $20 U.S. for each copy (plus $4 postage per order, not per book, in the U.S. Inquire for shipping cost to foreign countries). Alternatively, copies may be purchased from the author's website, and paid by credit card:

http://www.ggentile.com

The cover molds were grown
and photographed by the author.

International Standard Book Numbers (ISBN)
1-883056-35-7
978-1-883056-35-3

First Edition

Printed in the U.S.A.

Prologue

It happened in Australia.

It could have happened anywhere – in any one of a thousand places in the world. It could have happened in Denver or New York or London or Moscow. It could have happened in a government facility, in a large university, in a pharmaceutical laboratory, in a chemical factory, or under a baseball stadium.

With all the possibilities, a discovery as fundamental as this one almost had to be made some place, some time. But it happened in Australia.

The occurrence was not the result of a nuclear explosion, manmade environmental contamination, cosmic ray bombardment, or natural evolution. It was an accident, pure and simple.

The experiment was conducted not by a brilliant scientist, but by two small boys who inadvertently mixed unidentified chemicals in random quantities into an unknown solution to create . . .

But let me start at the beginning and go over the probable course of events step by step, the way I reconstructed them. The place: the southern portion of the Northern Territory, close enough to the center of the continent to be very sparsely populated, even by Aborigines.

The time: you may as well call it the present, although I am relating this narrative several months after the actual occurrence. It took that long for the initial horror of my ordeal to wear off. Call it the winter of 1964, although in Australia it was the middle of summer.

The scene: this itself is quite complicated.

The MacDonnell Range is a series of precipitous mountains near Alice Springs. Near the base of one slope there sits an aging schoolhouse that consists of a poorly equipped makeshift workshop that doubles as a miniature laboratory, and two classrooms on the opposite side of the hall. At one end of the hall is a rotting door to the outside; at the other end is a community room, small in comparison to the accepted notion of a community room, but large in that it is the biggest room within a radius of thirty miles.

The teacher is an underpaid English professor who volunteered his soul away to get where he is so he might teach some of the unfortunate natives the ways of civilization, while at the same time experimenting with extracts of local vegetation. He teaches everything from books, to which he adheres closely, except in his own specialty, which is biophysics. Through his political connections, he somehow manages to have supplies shipped periodically from Alice Springs.

In his spare time he works with herbs in what can be loosely termed as alchemy.

He is not strict with his pupils; instead he is compatible. He urges his students to come into his lab and watch, and even take part in, his offbeat chemical and biological experiments. In return for such fascination and fun, the students bring him rare herbs that only the Aborigines know how to find.

It was an extract from one of those medicinal herbs, and its combination with an organic airborne accelerant of unknown origin, that nearly brought humanity to its knees.

Of course there's a girl. Every story has a girl. But this girl was not the type who falls down and sprains her ankle at inconvenient times. Quite the contrary, I had a difficult time equaling her both physically and mentally. We got off to a rather bad start that quickly got worse through verbal sparring.

I remember vividly that day when it all began . . .

Chapter 1

The rain hammered against the Plexiglas port so hard that I could barely make out the signal lights on the madly vibrating wingtips. Water rolled down in rippling sheets that further obscured the view, adding a willowy, wavy effect to that produced by the huge raindrops that plinked against the window surface. The inset panes must have created their own slight eddy currents: how else to explain the lazy, careless raindrops that gravitated toward the bottom of the window and collected in a streamlined puddle in what I knew to be a better than five-hundred-mile-per-hour wind?

That the huge jet was descending toward the storm-swept runway at JFK, my many senses told me. The "fasten seatbelts" sign winked steadily overhead; the air was slowly clearing of bluish pungent mist as the cute and more-than-efficient stewardess reminded those errant smokers who were taking their last drag that all cigarettes must be extinguished during landing. My ears popped interminably as I worked my jaw to relieve the pain of reverse sinus blockage. The bitter taste of the last motion-sickness pill still lingered in my mouth, while the pill itself seemed to somehow have lodged in my throat where it was sure to do no good. Below, my stomach pitched and yawed in an attempt to bring the tasteless in-flight dinner to closer proximity of the stuck pill.

Outside the window, beyond the wingtip, was blackness.

As the jet ran afoul of an unusually strong downdraft, the cute and more-than-efficient stewardess suddenly plumped down on one knee, one hand steadying a last minute tray of glasses, the other hand gripping firmly the arm of my seat. Momentarily she crouched in the aisle, genuflecting toward the pilot's cabin as if she were an acolyte bearing offerings to the gods in return for a safe landing. I would have laughed if I had not

been so busy trying not to vomit: my dinner was determined to find that pill. I prayed that it didn't go past the
pill in its search.

Then we dropped from the swirling clouds and for a
brief instant I saw the tall spires of New York City
through the haze and beating rain. In the next instant
the plane banked, the stewardess rose lithely to her feet
and walked nonchalantly along the aisle, and I reached
for the barf bag on the back of the seat in front of me.
I had a vague impression of flashing lights, screeching
tires, and arrested motion as my body wrapped itself
around the seat belt. The added pressure was all that
was needed on my already churning stomach. My body
arched forward, vibrating rhythmically from the waist
up as I coughed like a fifty-caliber machine gun spew-
ing lead at an unseen enemy. What I had been vainly
trying not to accomplish for the past twenty minutes
finally came to pass: my recent dinner and the air sickness pill merged into one, amorphous mass – at the
bottom of the bag.

By the time I raised my head, the sound of applause
and cheering was resounding loudly in the fuselage of
the plane. Sheepishly I peered into the few surrounding
faces while a strange heat coursed along my spine and
raised the hair on the back of my neck. I felt my face
blushing. Finally I realized that my fellow passengers
were praising the pilot for his expertise in making a safe
landing, and that my raucous encore had had nothing
to do with their unbounded relief.

I wiped my mouth on a paper napkin and tucked it
into the dampening bag. Wordlessly I passed it to the
cute stewardess on her return trip from the front of the
plane. I expected a look of disdain, but she accepted the
offering with equanimity, holding it daintily by the top,
and marched toward the waste receptacle at the rear of
the plane. I longed for a drink of water to wash away
the foul taste in my mouth, but felt disinclined (or too
embarrassed) to ask for it.

As the plane neared the brilliantly lighted terminal,
I arose with everyone else despite forewarnings from

the intercom to the contrary, which bid passengers to remain seated until all motion had stopped. I hefted my bulging camera bag onto the seat and draped the strap over my shoulder. My briefcase was lighter, containing only papers and notes of my overseas trip.

Being a reporter and photographer (or photojournalist, the description that was used in the biz) had its elements of interest, one of which was travel. At least once a month during my two-year association with the *New York Script* I had been sent on assignment to obtain interviews with leading scientific personnel. As science writer I often ventured to other countries so as to bring the foreign element more into focus among smug American readers who believed that the only worthwhile advancements in science were being made in the United States. It was my intention to alert the public that other countries had made tremendous strides in certain areas of science, medicine, and invention, and that in many particular instances strode beyond American advances.

Of course, the job did have its drawbacks. I sometimes spent so much time away from home that it seriously hampered my social life – what there was of it. I found it difficult to get involved in local happenings or make any meaningful relationships with members of the opposite sex, since I never knew from one day to the next when I would be called away on assignment. I found it impossible to make a date for Saturday night any earlier than Friday afternoon, and even then sometimes my plans went awry.

On the other hand, the job often gave me an excuse to get away from situations and circumstances which had gotten out of control. It afforded me the luxury of absolute freedom and mobility that I might not otherwise have had if I had been clamped down to a simple desk job such as library research or rewrite. And besides, as meager as it was, the pay was better. All in all, I guess I had no reason to complain.

By now the aisle was full and the plane was jerking to a halt. My briefcase slipped out of my hand, bounced off the seat, and banged onto the floor, bursting open and spilling my copious notes in all directions. Several pages got trampled before I scooped them all together and managed to shove them into the briefcase helter-skelter.

I had spent practically the entire plane trip from Munich in sorting and rewriting my lengthy conversations with Dr. Wolfgang Brunholf, the Austrian brain physicist (or neurologist, in *his* biz). He had formulated some amazing concepts of the evolution of intelligence leading to human intellect. He theorized ways of detecting intelligence in animal life that was low enough on the evolutionary scale to be presumed to respond to external stimuli strictly by instinct. Sometimes the difference between intelligent motivation and instinctive reaction bordered a thin line. He even hinted that plants might have a kind of intelligence that was only different from ours, not necessarily absent.

Moving with the flood of people I made my way toward the front of the plane. I stepped out of line at the mid-plane kitchen and, since it was unattended, helped myself to a glass of water. I swirled it around inside my mouth, spat into the sink, and repeated the procedure twice more. I looked for some after dinner mints, but could not locate them. I lunged back into the crowd and crawled along toward the disembarkation ramp.

I smiled with an inhale at the head stewardess so she would not smell my treacherous breath. Rain pattered down on the thin sheet-metal that covered the moveable aluminum stairs, reminding me of the foul weather that was running rampant outside. The reminder struck me full in the face when I reached the bottom step and had to dash to the gateway in the open air. One degree colder and the rain would have been snow.

I purchased a package of mints on my way to baggage claim and threw half of them into my mouth, munching noisily as I walked. My camera bag kept jabbing me on the hip so I was constantly moving it in front where my too soft stomach could protect my bones from the constant hammering.

As usual my suitcase was among the last to arrive. It came out of the chute on its end and, as soon as it hit the downward sloping slide, it began to tumble. It ricocheted off a beige trunk and slid off the ramp, coming to rest on the smooth metal guardrail before reaching the circular conveyor belt, which displayed baggage to waiting passengers. A pair of red and white paisley boxer shorts protruded from the middle, trapped by one leg.

Indecorously I jumped across the circular conveyor belt to retrieve it. Upon my return to the ground I found myself in an argument with several baggage claim personnel for jumping across the conveyor belt. They contended that I should have waited until all the baggage was claimed and the belt was switched off. While I tried to hide my chagrin and argue for my rights, several nearby travelers began to snicker. I didn't think it was a particularly funny matter and I was both swept with rage at the attendants and embarrassed at those laughing clods who were having so much fun at my expense.

Although I was fairly positive that I had locked the suitcase after foreign customs had inspected it, and before handing it to airline authorities in Munich, the thought crossed my mind that it could have been rifled in transit. I steered myself into a corner next to a soda machine, sat the suitcase on its bottom, and looked at the combination lock. It was set on some random pattern, just like it would be if I had locked it and pushed the three dials with an errant finger. I set all three dials back to zero, then started moving the first digit.

I leaned back on my haunches, ran my other hand through my hair, then set the first digit back to zero. I stared at it for a moment, cursing silently. Thoughtfully, I rubbed my hand over my chin, feeling the coarse stubble of my four o'clock shadow. After several seconds I reached up and pulled my ear lobe. It was no use. I just couldn't remember the combination. Perhaps if I hadn't been so rattled by all the things that had happened in the last half hour it might have come to me. But in my present state of mind I was a complete blank.

I heard the sound of clanking change hitting the cams of the soda machine, followed by an insane cackle. Over the top of my suitcase I saw a pair of dirty feet casually encased in old leather sandals. As I elevated my eyes I saw American flag patches on the knees of a pair of faded dungarees, held up by a flowery leather belt with a big brass buckle that was a closed fist with the middle finger pointing up. Following in the direction of the brass rigid digit I saw that the creature was bare-chested, while over his shoulders was draped a loose-fitting leather vest. If I had not seen the flatness of the chest I would never have taken the face for that of a male. The features were soft and smooth, almost immature, and surrounded by brown curls of hair that reached the shoulders and lay there loose and scraggly. A leather hat was pulled down almost to the eyes, the cheeks were puffed with pink, and mascara was thick around the eyelids.

A surprisingly deep, sonorous voice called out from between thin, sensuous lips, "You better grab your shorts, man, they're trying to run away from you."

With that a can crashed into the bin, the creature scooped it up and did an about face, and sauntered away unconcernedly. A patch sewn to the seat of his pants, which was about on level with my line of sight, read in bold letters, "If you're looking here, you're perverted."

It was the last straw. I grabbed my suitcase and made a dash toward the taxi stand. Many smiling faces greeted me. I soon determined that they were looking first at my suitcase with its waving red and white pennant. I turned the suitcase around so that the hanging shorts fluttered behind. I steadfastly refused to see who or what was behind me. They could laugh and be damned.

As usual, the taxi stand was crowded. People huddled close for warmth, and stayed well out of the way of the pelting rain which was quickly becoming sleet with the dropping temperature. Normally I would have asked if anyone cared to share a ride to the City, as it was against company policy to take cabs alone. Every item entered on the expense account was gone over meticulously, and any squandering evoked a quick reprimand. It was well known all over town that the *New York Script* was in financial difficulties. Therefore, every penny was scrupulously counted (some said by Old Skinflint himself, our editor-in-chief, Mr. Ronald Gadfly) and any excesses met with serious challenge. I'm sure if it were possible he would have me washing dishes all the way across the Atlantic to help defray the cost of my plane fare.

Time was of the essence. It was afternoon already, and I had to go directly to the office to type my article so it could be set in Linotype this evening. Gadfly was expecting this story to appear in the next issue, which would hit the newsstands tomorrow in order to capture Monday morning commuters. Instead of waiting, I veered toward the limousine stand and, luckily, caught one just about to depart for downtown. The driver threw my suitcase on the roof rack while I jumped in with my camera bag and briefcase.

It wasn't until I was finally settled in my seat that I realized that I was shivering. Despite my warm clothing, specially packed for my trip to the Austrian mountains, the dampness went right through me. As we pulled out from under the vestibule I noticed that the sleet was beginning to accumulate into slush on the roads and sidewalks, making the surface slippery and treacherous.

I flashed a wan smile to the other occupants of the limousine. They looked like a pretty ordinary bunch: each was quiet, with his own thoughts or worries. The driver, of course, sounded like a continuous news broadcast on an FM station. The only useful information I gleaned from his rambling caterwaul was the fact that the bottom was going to fall out from under the thermometer and we were due for a lot of snow and extreme cold.

At this point I kept having thoughts of a hot shower and a warm bed. I was so tired and exhausted from my trip that I was in no mood for anything else. Most of all I dreaded having to report to the office to type the final copy of my story for tomorrow's issue. To make matters worse, there was no compensatory time allocated for extended work hours, so I would have to report tomorrow morning at eight o'clock to begin the normal week of routine overwork. I just kept reminding myself of that a hot shower and warm bed were waiting for me in my seldom-used apartment.

Being Sunday, traffic wasn't too bad, although thickening snow conditions slowed progress to a crawl. I was the first to leave the limousine, getting off where I could obtain the uptown train to the *Script* building. I gave the driver a healthy tip and said the hell with Old Skinflint, this guy had some hard driving in these conditions and his aimless rambling dialogue did have a quelling effect on me.

I still couldn't remember the combination to my suitcase and now, since it had spent the trip from the airport to the city on the roof, the boxer shorts were soaking wet and hanging like a soggy dish rag. I tried to forget the whole impossible episode as I trudged underground to the train station and concentrated on what I wanted to say in my article so I could write it quickly and go home.

When I stepped onto the sidewalk again (after a train ride that would take a book to explain – what was that combination?) the roads were full of quickly freezing slush, and the snowstorm was in full swing. Wide, fat snowflakes fell thick on the ground and were already piling into drifts. I pulled up my coat collar, drew my belt tight around my middle, and headed toward the ancient building that served as the headquarters of the *New York Script*.

A brass sign hung silently above the revolving doors, chipped and pitted with the patina of age. Years, decades of striving stood behind that lonesome, silent portal. Yet, if rumor were to become truth, its doors would soon be revolving no more. Despite large cut-backs in spending, a necessary cleavage in personnel, and my own astute articles, the *New York Script* was tottering on its last legs. I felt more concern for the death of this ancient periodical than I did about my own personal occupational loss, for in the two years I had worked here I had truly grown to love it. I have no doubt that if I had instead gone to work for the *Monday Morning Post* I would have felt these same feelings of nostalgia for that magazine. But my application had been turned down by their editor in favor of someone else with, I presume, more experience or better credentials, and I had had to search for employment elsewhere. So, I had found a home in the *Script* and, truth be told, they had not done badly by me, for if I had gone to work for the *Post* I would not have advanced to the position of chief science writer in such a short time.

But what was the sense of kidding myself. I was nothing more than a big frog in a small pond. Disgruntled writers working for the *Script* had an unusually short turnover period. Everyone knew that writers for the *Post* received more pay and greater prestige than the *Script* could – or would – offer.

Good, bad, or indifferent, I accepted my position with philosophic temper. Walking through those doors was almost like going home – more so, in fact, since I spent more time in this unadorned stone building than I did in my own apartment. Once safely ensconced in the small lobby I shook the snow off my shoulders. It was deliciously warm inside, and I paused a moment for my spine to stop shivering before continuing on to the elevator bank. My right arm was sore from carrying the suitcase so I switched hands, tucking the briefcase under my arm and swinging the camera bag in front. The switcheroo brought to my attention the fact that my exposed boxer shorts were still frozen in an attitude of attention. I ignored them.

A single gong alerted me that one of the elevators was about to go up. I charged into the waiting car a split second before the clashing steel doors slid shut, dropped my suitcase, and spun around to push the button for my floor. Meanwhile, the doors closed with a clang and too late I pulled back my suitcase. The protruding boxer shorts were snagged in the steel grip. I wrestled with them like a trainer pulling a bone from the mouth of a Doberman pinscher but there was no getting them loose until we reached the eighth floor. When the doors finally opened, the once red and white paisley shorts were torn and blackened by oil. I shook my head in solemn forbearance.

Sunday afternoons in the office were pretty slow. I counted on peace and quiet as a suitable atmosphere for writing. Most of the copy work had been done on Friday and all the normal people had the weekend off. My own desk was a mass of loose papers that had piled up during my ten-day absence. I swept everything aside and plopped my camera bag and briefcase over the dust swirls. My suitcase I leaned against the filing cabinet. After I wrestled out of my heavy overcoat and flung off my woolen scarf and hung them near the radiator to dry, I made a beeline for the coffee maker and started brewing a big pot of black lifesaver. While I waited for it to percolate I munched some crackers that I kept on hand in the drawer with the rubber bands and paper clips. Thoughtfully, I thumbed through my now disarrayed notes, sorting and sifting. But I steadfastly refused to do any typing until I had had my first two cups of brew.

The crackers filled a void in my empty and still upset stomach. Slowly, as my body warmed and my tension eased, my mind began to function and I rehashed my trip to Austria and my conversations with Dr. Brunholf. His advanced theories on intellectual growth and sensory stimulus were certainly revolutionary and quite extravagant, but not altogether unfounded. In fact, after considerable research on the subject in the Munich Library, I found that his conclusions became readily believable. If only I could capture that believability in my article it would go far toward dispelling the ridicule that he has been forced to endure.

Scientists were often like witch hunters: they did not suffer a heretic to live – at least, not in scientific circles. Brunholf was to neurophysiology what Velikovsky was to astronomy: beyond the pale but uncomfortably close to truth.

The bubbling of the coffee maker brought my thoughts around to more pertinent matters. I poured myself a cup of the steaming brew, added two lumps of sugar and one level spoonful of condensed milk that I opened for the occasion. Stirred quickly with some cold water from the fountain, I was able to drink down the first cupful in a flash, refill it, and do it over again. When I had downed the second cup and refilled the steaming vessel for a third time – without the cold water – I sat at my desk and again pondered over my notes.

As I scanned the many sheets of neat but cryptic handwriting, I reviewed Dr. Brunholf's theories in my mind. It would be necessary to use extensive quotes, for I could not otherwise hope to recapture his verve, his vivacity, his absolute sincerity. But for readability on a layman's level I would also have to put down my perception of his theories in my own words. It was important to the public that I not only report my findings fully for the intellectual subscriber, but that I understand it enough so I could restate it intelligently in a manner that was comprehensible to the lay reader. If I could do all that, and add my own flashes of insight, the article would be more penetrating, more polished, more expert than a straightforward regurgitation. That was what made the article mine and not Dr. Brunholf's. It was what I did to maintain my image as a crack science writer instead of a hack reporter.

With this determination in mind I sat down to maintain my reputation. I pulled the typewriter out of the drawer and set it on the desk blotter. I spread my copious notes in front of me and arranged them in their final order. Actually, the bare bones of the article were already written. Now it was a question of sifting and collating, smoothing and polishing, and – don't let the readers know this – adding fat to the meat so it was understandable to the public at large.

I ran my fingers over the keys to get the feel of the ivory, so to speak. Inserting two sheets of paper with a sheet of carbon paper in between, I typed a couple paragraphs of quotation from my handwritten notes. Then, tearing out the rough draft and inserting a fresh triplet, I began to develop an article.

Page after page ripped through the typewriter. The article seemed to be writing itself as my subconscious mind collaborated with the keyboard. Apparently, my hours of speculation and days of research in the Munich library had been well spent, for concepts seemed clearer in my mind now than I remembered them to be when Dr. Brunholf was explaining them to me. The questions that had bothered me before now vaulted forth in startling clarity, the answers seemed obvious, and the deepest thoughts no longer seemed profound.

"Intelligence quotient" is a term that is used to describe the brainpower of the subject being tested. The higher the quotient, the smarter the individual, according to traditional thought.

Dr. Wolfgang Brunholf disagrees. According to him, IQ tests are biased not only by cultural imperatives and examination processes, but by the very nature of the species that devises the tests. Other animals have intelligence that equals or exceeds that of human beings, but their kind of intelligence cannot be determined by written or oral exams. They must be tested in relation to the environment in which they live. Tigers, for example, are extremely intelligent, made evident by their position at the top of the food chain. Yet their *kind* of intelligence cannot be tested by current methodologies.

Even IQ tests for school-age children and college-bound adults are poorly designed. By and large they test only two facets of intelligence: memorization skills, and the disgorgement of information that they have been fed. Ask a person questions for which he has not been prepared, and he will fail miserably to obtain a so-called "passing score."

Many slow learners achieve remarkable success in the worlds of business and finance. Many genius types hold menial jobs. The obvious inference from these mental incongruities is that there are facets of so-called human intelligence that go unrecognized and unrecorded: initiative, dedication, concentration, and maturity, to name a few. These facets can be measured after the fact, but not tested beforehand.

Human intelligence originates from the neocortex of the cerebrum. Nonhuman mammalian intelligence resides largely in the cerebellum: the so-called "hind brain." Animals that occupy "lower" places on the evolutionary scale survive by instinct: a catchall word or naming convention for a mechanism that is not understood. Yet animals such as insects survive quite well and in abundance.

Dr. Brunholf believes that there are kinds of intelligence that are incomprehensible to Homo sapiens, just as there are colors that man cannot perceive. Even plants exhibit intelligence that most scientists scoff at.

And so it went. When I copy-edited the first page, I deleted the repetitious occurrences of "so-called," and changed the final sentence so it did not end in a preposition: "Even plants exhibit intelligence at which most scientists scoff." The stiffened formality was better than the colloquial equivalent. Articles in the *New York Script* did not have to be overly erudite, but neither could they afford to be too familiar, else the magazine would lose its credibility with readers who associated themselves with the intelligentsia.

I also changed "Dr. Brunholf believes" to "Dr. Brunholf thinks." I did this not because "believes" implied religious connotations, but because "thinks" suggested cogitation and scientific analysis.

I added a sidebar explanation of the workings of the brain: how a neuron (or nerve cell) extended dendrites to the axons of adjacent cells, and transmitted electro-chemical signals (or impulses) that were interpreted and passed on to other cells. This was not only the process of conscious thought, but the means of controlling the autonomic nervous system.

Three cups of coffee, two packages of crackers, and a candy cane left over from Christmas later, I whipped the final page out of the platen. By this time I was so hot – literally as well as literarily – that my vest had been divested, my tie untied, my shirt unbuttoned, and my armpits sweated. The article, I felt, was definitely a success. I danced exuberantly around the wastepaper basket, separated the carbon paper from the sheets, and dropped the carbon paper into the circular file.

I was stuffing the carbon copy into my briefcase to take home when the telephone rang.

Chapter 2

As I scooped up the instrument and squeaked a tart "Hello" into the mouthpiece from a surprisingly dry throat, an explosion of noise assaulted my left ear.

"Baker!" the phone shouted at me. "What the hell is going on. I called the airport and found that your plane had arrived hours ago. You should have called me right away. Now I want an explanation. Where the hell have you been?"

There was no mistaking the trombonelike voice of Editor-in-Chief Ronald Gadfly. His thundering vociferations had been known to shake desks off their legs and reporters off their feet. From his dense overtones I got the impression that I had done something wrong, or at least committed a crime. Actually, this was nothing more than his normal tone of voice and his every day attitude. Nevertheless, I was cowed into submission.

"Why, Austria, sir," I uttered plaintively, barely loud enough to actuate the diaphragm of the phone.

"I know you were in Austria, confound it," he screamed. "What I mean is, why didn't you report to me as soon as you got in? Why didn't you follow my directions?"

"Well, sir, there really didn't seem to be any reason to call until I had written my article, so I thought . . . uh . . . I thought . . . uh . . . what directions were those, sir?" I was more confused than dismayed. His voice, however, did not lack for luster.

"Confound it, the note I left on your desk. It's as big as life."

I stalled for time as I moved things around on the desktop, retrieving books, papers, and miscellaneous items that I had so hastily pushed aside when I had first arrived at the office. "Oh," I said, in frank surprise, "that note."

"Yes, *that* note," he mimicked.

Then I saw it. It was written on oversized sketching

paper in black indelible ink, and it was indeed as big as *Life*: the title on the cover of that magazine, that is. In seventy-two-point bold it loudly proclaimed, "BAKER – CALL ME IMMEDIATELY UPON ARRIVAL – IMPORTANT – R.G."

"Well, sir, you see," I stammered, "I figured it would wait a little while until I could get my things together before I reported in. You see, I only had to touch up the article in a few places and tweak a few passages, and then I thought . . . "

"Confound it, stop thinking and stop talking and get yourself up here before I send someone down there to kick you up here. I'll give you two minutes and if you're not here by then I'll have your check drawn up."

A crash in my ear told me that the conversation had been terminated. I was not particularly worried about the threat of termination, however. Invariably I got fired at least once a month; so, in fact, did everyone else at the *Script*. It was even chided in the inner circles that a month passed without being fired was an unproductive month. So Gadfly's comment wasn't anything to worry about. Nevertheless, it would not do well to provoke Old Skinflint – at least, not until I knew what was going on.

Without waiting for the elevator I charged into the fire tower and ran up five flights of stairs, emerging in the hall at the opposite end of the editor's suite. Sprinting down the corridor I breezed past row after row of empty desks and side offices – the secretaries had weekends off, and how I envied their banker's hours. I halted only momentarily at the double glass doors that separated the spacious administrative offices from the common work area, and kept on running.

I exploded through the oak door into Gadfly's office without knocking, stood there dramatically for a moment before commenting, "I came as fast as I could, sir." I sucked in my breath in great gasps, feigning hyperventilation. Due to my constant jogging I wasn't really out of breath. I strode up to his desk, proffering my completed manuscript, and stood there breathing hard, my shirt top unbuttoned, my hair disheveled.

As our gazes locked onto each other I studied his hard-set features. A mop of thin, graying hair swept back from a round, creased forehead and was trimmed neatly above the ears. Deep lines spread outward from hypnotic blue eyes and surrounded a pudgy, ruddy face like lines of magnetic force that kept iron filings circled around lodestone. The bulbous nose was the result of too many punches received when he was just a hotheaded and hard-boiled cub reporter roaming the streets for a story. The twitching mouth was a sign of nervous energy that was never quite drained. Bags beneath the eyes and a stout neck crammed into a too tight collar attested to overwork and underfasting. His body was as round as his face, and one would never suspect that it could react as quickly and as effortlessly as it did. If one were to describe Ronald Gadfly in a word, that word would be energetic. And now this bouncing ball of energy sprang into action.

My histrionics were all for naught. Gadfly snatched the papers from me with one hand while reaching for the phone with the other. He seemed not to be impressed at all by my prompt arrival, despite the fact that only twenty seconds had passed since the time he hung up on me and the time I arrived at his office. A mere third of a minute to dash up five stories and traverse the length of the building was taxing even my great speed. I was instantly deflated. Gadfly proceeded as if this kind of action was expected.

"Thompson," he barked into the phone while rising to his feet and pushing back his overstuffed leather chair. "I'm sending down the Brunholf article. Have it Linotyped right away and get the presses rolling as soon as possible."

He hung up without waiting for a reply. In three steps he was standing by the pneumatic tube station. He rolled up my manuscript and shoved it into a clear plastic container, snapped on the lid, and shoved it into the tube opening marked "LINO." It seemed that as soon as the tube disappeared with a swish of air, it was forgotten.

"Baker," he shouted, on his way back to his desk, "I've got your next assignment for you."

Gadfly sat in his sumptuous chair and began rifling through some papers, almost abstractedly, while continuing to talk. "Have you ever heard of Professor Wayward? Professor Warren Adolphus Wayward?" Without giving me time to answer, he went on, "Probably not. In any case, Wayward is a biochemical physicist or some such thing with a string of letters behind his name that would make up three alphabets. He belongs to virtually every scientific society in England and most of the ones abroad. He holds emeritus professorships in two colleges and acts as a consultant to several conservation boards throughout the world. Ah, here it is."

Gadfly separated a file folder from the bric-a-brac on his desk and shoved it across to me. I took it mechanically. Inside was a thick sheaf of notes, bio sheets, newspaper clippings, and magazine articles, topped by an eight-by-ten black-and-white glossy print: a head and shoulders photograph of a middle-aged man in a business suit.

I studied the portrait carefully. He was a distinguished-looking aristocrat with the mien of a stereotypic nineteenth-century scientist. A mane of straight, snow-white hair swept back from a prominent forehead that was almost the same bleached color. That the forehead was broad and angular rather than sloping put his features into perspective on the same scale. Bright, piercing, slate gray eyes stared enigmatically into the lens in such a way that they followed wherever I looked. The nose was pointed but not quite aquiline, the lips thin and drawn, the cheeks shallow, the chin strong and obstructing. His upper lip was thickly crowned with a bushy white mustache. His skin was smooth and pale.

"I've never heard of him," I finally intoned.

"Of course not. He's been out of circulation for years. He had been doing secret government research for the Ministry of Defence at Woomera – god knows doing what: even *I* couldn't find out. He had his own laboratory, his own assistants, and everything he asked for, including a monumental salary.

"Then suddenly, the government lets him go. The cover story is that bureaucratic intervention prevented him from following his own inclinations in certain areas of study. I don't believe it. He said he would rather deal with life than with death (which is as close as he came to admitting to the type of work he was engaged in, and, I might add, he was severely censured for that statement). He said he wanted to extend his researches in new directions, to use his expertise for the good of all mankind, and all that other humanitarian crap. Anyway, he moved into the wilderness and went into seclusion. Recently it was announced that he was using radiological bombardment techniques to stimulate growth patterns that were genetically transmittable in plants, the main idea being to provide more food more economically. I guess he's trying to save the millions of poor and hungry bastards the world already has too many of so they can give birth to more, or some such nonsense. And anyway, how do you run a radiology lab in the middle of the wilderness?"

I ignored Gadfly's opinionated views. What interested me was the characterization of a man who so glibly chided the British government. Scientists were usually forthright, inherently intelligent, bold in their views, but never antiestablishment, particularly when their jobs, or grants, depended on a certain amount of kowtowing.

Under the portrait of Professor Wayward, I leafed through the scrapbook material. Reading, however, was only a ploy so that I could avoid that penetrating gaze with which Gadfly transfixed his listeners. I started to talk once but was instantly cut short.

"I've been working on Wayward's history for six months but the Ministry of Defence is very oblique about his old lines of research, about what progress he made, if any, even about his present whereabouts. Finally, through a little ambassadorial pressure, I found out where he was working and wrote to him personally requesting an interview. Surprisingly enough, he had no objections. In fact, he seemed surprised that anyone was interested enough in his work to want to interview him.

"The major stumbling block had been the British government. All Wayward's projects were under top-secret security wraps, with little chance of circumventing protocol. There is a strict confidential quarantine on all aspects of his prior research, despite the fact that it is ancient history by now. Now I did manage to come up with this, from sources which I can't reveal."

Gadfly rooted through some more debris on his desk and came up with another folder. He handed it to me and continued his monologue. (How he managed to work with such proficiency in such complete disarray, I'll never know.)

"This represents all the background material I've been able to come up with. What you see there cost a lot to get, in both money and favors. It gives some official data on Wayward's life, his experiments, his successes, his involvement with the scientific community, and his expertise in the field of radiological growth. Don't ever, for any reason whatsoever let those papers fall into British military hands or we'll both end up cleaning toilets in a Tibetan monastery for the rest of our lives. Especially I don't want our competition to see it."

"Competition," I said, for the first time returning his steadfast gaze. "Who, or what, is our competition?"

"I believe, too, that the *Post* is also sending out a writer, name of Crawford, related somehow, I gather, to the editor-in-chief, John Nolan Crawford."

I certainly knew that J.N. Crawford was the editor of the *Post*, since I had applied to him personally for a job. But the name was hauntingly familiar in another vein. I seem to recall reading an article with a Crawford byline about life in the nuclear world, but with different initials. I remember, too, that the article was not written very well – although the facts appeared to be accurate. One way of keeping tabs on your competition was to read their material.

"How did they manage to line up an interview with this guy Wayward at the same time you arranged ours?"

"Seems as if the higher-ups – " He pointed a finger at the ceiling to indicate the senior management staff that occupied the upper floors – "have been conducting some kind of secret confabulations with the higher-ups of the *Post*. I suspect that the contretemps between Nolan and me has rankled some fur. I think someone tipped our hand.

"Crawford and I, while not exactly friendly, respect each other professionally, and, although I don't want to do anything underhanded, I want to beat him to the punch on this scoop. He must have done some spying to keep so well informed about us. I want to make sure that he doesn't profit by his shenanigans by scooping the *Post* on the Professor Wayward story."

Gadfly shook his head. "I sure wish I knew what was going on upstairs." Again he pointed his finger at the ceiling. "There's been a leak in all my negotiating, and somehow Crawford found out about it. In any case, I want you to be on the lookout for Crawford's nepotist. We've got to beat Crawford and the *Post* to the punch or this story won't amount to a hill of beans with the folks upstairs." For the third time he pointed at the ceiling. "That's why I want you to get on this assignment right away."

The whole affair had an air of subterfuge about it. I was incited by the idea of tracking down a story about the mysterious Professor Wayward, but did not like the cloak of clandestinity with which it was encumbered. There were too many hidden nuances that smacked of trouble. Tangling with another reporter from a rival magazine was bad enough, but defying British governmental edicts could seriously curtail my professional standing.

And besides all this, I had personal priorities to attend to. First and foremost was a certain female I had been dating rather sporadically over the past several months. If I expected to make any headway at all into her affections, I was going to have to be a little more attentive to her needs and more consistent with our social engagements. Will all my irregular field trips, it was difficult to maintain any kind of schedule. She would probably want to see me tonight, but I would have to beg off so I could get some sleep.

Then there was the upcoming race. In two weeks I was entered to run in the annual twenty-six mile New York Post Collegiate Marathon. This year I was heralded as the most likely winner, since I had come in runner-up last year, and my chief competitor was out of the action with a broken leg, the result of a skiing accident. Even in Austria I had found time to go running practically every day, but now it was imperative that I go into a strict routine of training if I wanted to walk away – or rather, run away – with the gold cup. What I needed now and for the next two weeks was more sleep and a lighter workload. Running through thick London fog was not my idea of ideal training conditions.

Still, if I could get a good night's sleep, fly to England in the morning, I could have three days to work in interview sessions with Professor Wayward, fly home on Friday, transcribe my tapes on Saturday, write the article on Sunday, and still be able to take Julie to that opera on Sunday night with the tickets that I bought a month ago. But then I would need to have a week of good sleep and plenty of exercise in order to get ready for the marathon.

It was possible to fit it all in. Just barely.

I must have had a look of consternation on my face, for Gadfly launched into another recitation. "I may as well tell you that there's another fly in the ointment, one that you've probably already heard about. Word is passing around that the *Script* is in deep financial trouble. Despite the quality of our production the competition is stiff. Circulation has been going down steadily for the past six months and unless a miracle happens we may not survive another season. We've got some representatives on the Board of Directors who would be happy to see this operation shut down and sold for its assets, or possibly sold to another syndicate. I don't have to tell you that in either eventuality we'll both go down the tubes. What we need is more time to show them that the drop in circulation is only temporary – a seasonal event that will reverse itself in due course. We need something strong and exciting to make our magazine stand out above the rest. I believe that if we can scoop the Professor Wayward story we can accomplish this. And I'm depending on you to do it."

That was a heavy statement. At first I thought Gadfly was using his confidential technique. By getting real buddy-buddy with me and letting me in on things, it was supposed to make me feel honored to know that I had his confidence. But as I looked deep into his penetrating blue eyes I saw that they were touched with concern and compassion.

To Ronald Gadfly, the *New York Script* represented the sum total of his achievements in life, and as such was as dear to him as life itself. He had striven all his life to get in the top dog position where he could direct policy, call the shots as he saw fit, steer the course of the magazine his own way. He was so enmeshed in the *Script* that he *was* the *Script*. And it could all vanish for him overnight. Sure, he had been a cheap miser: cutting incidental expenses, laying off drones, slashing expense accounts, putting the burden of his worries on us. But that was his job. He had to see that his magazine accomplished what its readers and backers wanted. He wanted it to succeed. He needed it to succeed. For if it didn't succeed, Ronald "Old Skinflint" Gadfly would blame himself for its failure. He stood to lose more than his job – he stood to lose his life and everything it stood for.

I was more than half convinced that I should accept the assignment – aside from the fact that I probably didn't have any choice in the matter. But I had to try just one more time to get out of it gracefully.

"What about Rogers or Awlsly?" I asked. "Why don't you give this assignment to one of them?"

"For one thing neither of them has your experience. And besides, neither of them are available at the moment."

I'll just bet they're not, I thought. Rogers was probably out on an all-weekend bender and wouldn't be conscious until sometime Monday afternoon, even though he'd stagger into the office Monday morning reeking with alcohol. And as for Awlsly, well, he was probably shacking up with some hussy he picked up over the weekend and has his phone off the hook.

"What do you have, staff infection?"

"Don't start with me."

"But you know I've got the marathon coming up." I decided to try one last time.

"From the way you were breathing when you came in here you'll never make it to the starting line. Besides, you'll be back in plenty of time for that race. Look, Tim, let's call a spade a spade. There's nobody on the staff of the *Script* who can handle this job like you can."

I started to worry when he called me by my first name. I felt like a piece of warm toast because I knew I was about to get buttered.

"You didn't go from cub reporter to chief science writer in two years by accident – you earned it. You're the best we have. Do you think if anyone else came in here with an article I would have sent it to the presses without so much as looking at it, as I just did with yours?"

I was beginning to weaken. He was hitting me in my one weak spot – my vanity. Then again, everything he said was true.

"Rogers and Awlsly are good writers, sure, but they're not reliable. I can't afford to have this assignment bungled. How can I trust either one of them on an assignment of this importance when one may drink himself into oblivion halfway through the job and the other might run away with the first whore he runs into? And besides that, they are only reporters. You are an author. You understand what you're writing about, you can put feeling into an article as well as knowledge and understanding. You can make the readers feel compassion for the necessity of research as well as making them understand the science behind it. You're much more than a reporter, Tim, and I need whatever it is you've got on this assignment. I need *you*; *we* need you. This whole publishing house is on the brink of collapse and you may be able to save it."

This was a historic moment in the life of the *Script*, for Gadfly, Old Skinflint himself, was as close to tears as he was ever likely to get. His impassioned plea was both honest and sincere. For the first time in my long association with him I felt that I finally knew him, because he had allowed me to enter his secret world, he had shown me his innermost feelings, he had bared his soul to me. To reject him at this crucial time would not only have crushed him, it would have closed forever the bond of friendship that he was offering to me. I melted like a wax candle left in hot sunlight.

"All right," I said, nodding affirmatively. "I'll take the assignment. Now give me the rundown on this guy Wayward."

His face brightened up like that of a child who had just been given a box of chocolates, and told that he could eat as much as he wanted. "Thanks, Tim, I knew I could count on you. As for the lowdown on Wayward, everything you need to know is in those folders." Gadfly glanced at his watch. "Now you'd better get a move on. Your plane leaves in an hour."

"*An hour*," I screamed. "I couldn't possibly leave in an hour. Even if I didn't have to go home to shower, change, and pick up clean clothes, it would still take longer than that to reach the airport. With this storm outside, traffic will be at a standstill. Besides, I need some sleep."

"It's a thirty-hour flight so sleep on the plane," he replied, feeling through all the pockets of his jacket and vest. "Why do you think I left that message for you to call me right away. You would have had plenty of time if you had followed instructions. Now listen, I've made a special deal with the airline. They gave me this ticket at half price in exchange for an ad in next week's issue. It's validated for this afternoon's flight. And I wasn't kidding when I said that the *Post* has this Crawford chap hot on your tail. If you want to get the jump on him you have to make this flight."

Rationally, I knew that it was too late to back out, but in the back of my mind I hoped fervently that I could miss the plane and at least get a good night's sleep before embarking on this venture. But at that moment Gadfly yelled, "Here it is." He pulled out an itinerary envelope from his inside vest pocket. He handed it to me with one hand and reached for the phone with other.

I thought I was returning to Austria until I noticed two extra letters in the middle of the word. The "al" made half a world of difference.

"*Australia!*" I shouted. I looked through the itinerary and saw that the plane was stopping at Los Angeles, Honolulu, and ending at Darwin. "There must be some mistake. This ticket is for Australia."

"That's right," he said calmly, as if we flew to Australia every day of the week. Suddenly I was wishing for the blustery skies and foggy shores of merry old England. "And remember, Professor Wayward's laboratory is at some outpost in the wilderness near Alice Springs. Apparently he went walkabout and settled in an obscure Aborigine village where he works as a teacher in a one-room school. But there is no way of booking that part of the journey from here. You'll have to do that on your own when you get to Darwin, but whatever you do, please try to keep the expenses down.

"But I can't go to Australia. Not now."

"Of course not now. Your cab won't be here for five more minutes. Your suitcase is already packed from your most recent trip. And remember that you only have to tip the driver ten percent." Old Skinflint was himself again. He had come around the desk and was ushering me toward the door.

"But what do I do when I get there? How do I know how to contact him?"

"There is a full set of instructions in that folder. You'll have to use contacts in Alice Springs to locate Wayward, and utilize local transportation. Bring back this piece as soon as you can. Do the rewrites and final touch-up on the plane. I'll be waiting for it. Good luck, boy. From now on you're on your own."

Then I was outside the office and the door was closed. I turned around and for a moment stared at the oaken portal, thinking viciously how I would like to kick it in. I knew when I had been bested. There was nothing to do now but go on with it. Somehow I had gotten myself into another undesirable position.

Someday I'm going to stand up to Old Skinflint and lay down the law.

Someday. But not today.

Chapter 3

The cab was waiting for me as I passed through the revolving doors of the *Script* building. The car idled softly at the curb, a puff of gray, billowy smoke streaming from the exhaust pipe, while the driver jumped from foot to foot and pounded his gloved hands together, striving for warmth in the midst of a blizzard. I pulled up my collar and hunched my shoulders in a vain attempt to keep the cold snow from spiraling down my neck. There was a bitter wind blowing and I could feel thousands of pinpoints of iciness penetrating my clothing, could feel the numbness settling into my exposed hands and face as my circulation lost ground to convection.

The cabbie held out shaking hands for my suitcase as I waded through six inches of fluffy snow. I stood by momentarily while he tried to get the trunk open, but the earlier freezing rain had done its work and the lock was frozen stuck. He finally managed to get the key into the lock, but the tumblers would not turn. He gave up in exasperation and said that he would have to tie the suitcase to the trunk rack. I grunted an assent through clenched teeth and abandoned him to his task.

As I settled into the back seat I heard the loud drone of the heater going full blast: the inside of the cab was deliciously warm. I absorbed the heat like a failing solar cell. With my aging handkerchief I wiped my tangled hair and the back of my neck where the snow was beginning to melt down my collar. Next to me the camera bag and briefcase dripped onto the faded upholstery.

The front door opened with a blast of swirling snow as the cabbie squeezed his rotund body behind the steering wheel. With gesticulating hands and colloquial imprecations he vented his anger about the weather, the traffic, and the political situation concerning road clearing and emergency snow removal. As we pulled away from the curb into the snowbound traffic, I passed him a five-dollar bill and mumbled something about having a lot of work to do before we got to the air-port. I felt the gravity of the assignment and my peace of mind warranted such an extravagance. To further dissuade him from conversation and in order to focus my thoughts, I opened my briefcase and buried my face in the voluminous notes that I had taken on my previous assignment – notes which I had not had time to sort through and leave behind.

I still had much to think about concerning my interview with Dr. Brunholf. The article, I felt, was a success. But there were many things I wish I could have written that I neither had time for, nor the readership. Although I had given a good overview of Dr. Brunholf's ideas, there was really much more that I had left out. I was seriously considering doing a more in depth study and having it published in one of the scientific journals. Although freelancing was not actually condemned, as long as I was not guilty of conflict of interest, Gadfly would not object if I did some freelancing; not much, anyway. Of course, what he considered conflict of interest and what I considered conflict of interest might be two different conflicts of interest. But I was sure that I could convince him to see my conflicting point of view.

I had already gotten permission from Dr. Brunholf to do another article, as he was very eager to have his thoughts projected in America where they were likely to be picked up by enterprising organizations with funds more readily available for pure research. He was quite content to be a steppingstone toward acceptance of his theories, rather than to be sole possessor without the wherewithal to proceed in his investigations.

I had enough background information to do a thorough and very professional article on par with any that was written by a competent scientist in the field. My own education especially suited me to the task. My college major had been general science – for some silly reason that I can no longer fathom, I wanted to become a high school science teacher – with creative writing as a minor. When I decided that I didn't want to put in the additional time to earn my masters degree – the only way to land a teaching job – I relied on my writing ability to earn my living.

A year in advertising forced me to the conclusion that there was more to writing than dreaming up catchy phrases for cookies, and creating inane comedy situations for the latest dishwashing detergent. That was what prompted me to respond to several solicitations for science writing. When *Scientific American* returned my resumé with deep regrets, I immediately shot it off to the *Post* where it at least put me in line for an interview. Crawford wanted someone older and more experienced (or so he said). At the *Script*, Gadfly was looking for someone young and energetic: a bill that I filled enthusiastically.

I accepted the position of cub science writer, and was mildly surprised to find that the beginning salary paid more than my position in advertising. And, of course, anything pays more than teaching – and offers fewer headaches. Now the name Crawford popped up in my mind. The more I thought about it, the more it bothered me. Then it hit me like a ton of bricks. Crawford was the one who had gotten the job at the *Post* that I had applied for two years ago. Once again I reminisced in rancor about the *Monday Morning Post*. Soon after I had taken my position at the *Script*, I read an article in the *Post* written by Crawford. It was poorly worded, discordant of thought, and bereft of insight. I had been outraged at the thought that some moron had landed the job which I so rightly deserved. The name appeared only sporadically after that. Then, after about six months, the Crawford byline was no longer used. By that time I was pretty well entrenched in the *Script* and decided to either sink or swim with that periodical rather than reapply with the *Post*.

Now, after two years, I found that Gadfly knew all along that there was a blood relationship between the editor-in-chief of the *Post* and the mag's newly hired science writer. I had lost the job not through my lack of credentials, but through family affiliation.

I reveled in the irony of the present situation, for this assignment not only pitted me against a competitive magazine which had at one time rejected me, but also threw me in the ring with my personal rival and antithesis. I was being given the opportunity to redeem not only the *Script*, but myself as well. This was a battle that would be fought on two levels. First, there was the *Script* against the *Post*, which meant Gadfly against Crawford (the editor). Second, there was I (the verb to be takes the predicate nominative) against the bungling reporter who had undeservedly acquired my job, Crawford the writer. My enthusiasm suddenly waxed as I steeled myself for the coming fight with all the vengeance of a wild nemesis.

My eyes danced with glee as I pulled the plane ticket from my vest pocket and viewed it in a new and different light. Old Skinflint was still cutting corners and pinching pennies, but I loved him for it. My elation did not end until I thumbed through the itinerary and discovered that the fare that Gadfly had weaseled from the airlines for half price in return for an ad in next week's issue was only a one-way ticket!

The cab suddenly slewed, skidding madly on the snow-covered street. I slid hard against the door, receiving the window crank in the gut and spilling the papers from my briefcase onto the grubby floor. The cabbie muttered something about other drivers, and by the time I collected myself and looked out the window, we were cruising along slowly as if nothing had happened. I glanced out the back window to check on my suitcase. It was still tied securely in place on the top of the trunk lid, with my stiffened boxer shorts protruding from one end, and was now completely covered with several inches of snow. I thought dimly of those paisley boxer shorts. They had fared worse than I had during my interview with Gadfly, for they had been so close to the radiator that by the time I had returned to gather my things for my hasty departure, the water had dried out and they had been singed black by the heat. I shook my head and thought about more pertinent matters.

The most important thing on my mind was that I was going to have to cancel that date with Julie. I hadn't had time to call her from the office, so I would have to try to get in touch with her from the airport as soon as I checked in. It was a conversation that I dreaded, for she was not the most understanding woman in the world. The only way I could think to appease her was to offer to take her somewhere more extravagant than the opera to which I had intended taking her. It might make her feel better, but it would not help my ailing pocketbook. I had already paid for the opera tickets, so that would have to be chalked up as a business expense – that is, money that I spent to keep my job until the time of liquidation.

As for the marathon, if I could find some time to train while I was in Australia, I might be in fairly decent condition by the time I returned to the States. If the interviewing and background research was not overly exhausting, I might be able to get a fair amount of rest during my leisure hours, and jog some half marathons in the mornings. If I could get the assignment wrapped up in short order, I should be able to plan a return flight in time to make the race. There were a lot of big ifs, and they were all predicated on the premise that I did nothing but sleep, train, and work, which left no time at all to see the sights. After all, how often did I expect to travel to Australia, all expenses paid (or most expenses paid) by Old Skinflint? That was assuming, of course, that another plane ticket would be made available to me for my return. I left both questions unanswered in my mind.

I felt my ears. They had been pretty cold in the snow, but were now warming up. I thought I had better take good care of them, for this assignment looked like it was going to be played by them.

I glanced nervously at my watch. If strict adherence to scheduling had been maintained, my plane had left twenty minutes ago. As if on key, a huge, lumbering jet launched itself overhead in the slackening snow, vault-

ing into the grayish-white sky that was still filled with
falling, swirling snow. If it were my plane, then all my
mental gymnastics had been for naught. Laconically I
forced myself to relax as the cab churned through the
eastbound traffic toward the sprawling buildings and
intersecting runways of JFK – newly named after last
year's presidential assassination of John Fitzgerald
Kennedy.

By the time we reached the terminal entrance I was
forty minutes late. It took another ten minutes once we
were off the Southern State Parkway before we coasted
to a stop in front of the doorway to the Eastern Airlines
ticket office. The snow was thick on the roads but coming down more slowly, and I could see snowplows working hard to clear the runways, while several lumbering
jets idled their engines on the approach paths. I paid off
the cabbie inside the warm car, tipping him in accordance with the prevailing road conditions and the expe-
diency of the trip. Then he stepped out to hand me my
luggage. The straps were frozen and he had to wrestle
with the snaps to free them. Three inches of fresh snow
had fallen during our ride but it was soft and fluffy and
most of it fell off when I grabbed the suitcase by the
handle. What remained was a coating of ice that covered the upward-facing portion of leather. Mumbling a
sharp thanks, I ran as quickly as I dared in the slippery
conditions for the security of the swinging doors.
Inside, people were milling by the hundreds, like stalled
cattle at a deep ford, and about as directional. Apparently many flights had been either delayed or cancelled,
and the travelers were piling up waiting for instructions. I weaved through the roiling mob, running whenever I got a few feet of clearance.

People kept staring at me but I paid them no mind.
I just kept hoping that my flight had been delayed. I
might miss the flight anyway, but it would not be for
lack of trying.

The floor was glazed with water and dragged-in
snow. Normally I'm pretty stable on my feet, but now
conditions warranted extreme caution. Several times,

despite my care, my feet skidded out from under me. Once I managed to save myself by hugging a fat porter who huffed in obvious consternation; another time I sat heavily on a convenient trunk, then stood up nonchalantly and continued on my way as if nothing untoward had occurred.

Then, above the clamor of the crowd, I heard an announcement over the loud speaker: "Eastern Airlines Flight 287 for Los Angeles is now boarding at Gate Seven." The rest was garbled by the screaming crescendo of humanity but I had heard enough already. For 287 was my flight!

Heedlessly I increased my speed. I knew from past experience that if I could reach the gate and turn in my ticket before they pulled away the stairs and closed the door, they would let me board the plane. In my bungling manner, I had missed – and almost missed – a number of previous flights.

I saw the sign ahead for Gate Seven and thought I was almost there. But when I rounded the corner I saw that the gate's proximity was merely an optical illusion, for there stretched before me a long corridor that angled down until in the distance the ceiling seemed to converge with the floor. This was also the corridor for Gates One, Three, Five, and Two, Four, and Six. My gate was at the farthest end of the corridor.

My breath was under control but my arms were beginning to ache from the weight of the suitcase. I kept changing it from hand to hand, while I controlled the swing of the other two cases around my neck with my free hand.

Down that unending corridor I ran, dodging people left and right, caroming off a support column, tripping over a stray duffel bag.

At last I saw it. A herd of people was congregated in front of Gate Seven, surging to get through the single portal. The ticket counter was deserted except for a straggler or two. Outside a low railing, visitors were waving their last goodbyes to friends and relatives.

The loudspeakers were again blaring, "Eastern Airlines Flight 287 for Los Angeles . . . ”

I started to decelerate, too late realizing that the combined snow and water on the highly waxed floor was not conducive to rapid stopping. As I slowed my pace my shoes began to slip on the polished linoleum and I had to keep up a modicum of speed to prevent myself from tipping over. I soon saw that my forward momentum would carry me past the ticket counter and possibly into the crowd

“ . . . is now boarding at Gate Seven . . . ”

Without realizing what had happened, I felt a sharp slap on my rump that rattled my teeth. The fluorescent lights on the ceiling shone down directly into my eyes. I fell in a heap of tangled straps and baggage; my suitcase bounded off the floor and lodged against a chair.

“ . . . but will be delayed for at least twenty minutes while the runways are cleared and the wings are de-iced. We apologize for the inconvenience.”

The last part of the message – the part that I had missed before – seemed to snigger at me. The first thing I did was to inspect my camera bag to make sure that all my lenses were intact; they were. After that I inspected my body for damage. It had fared well, despite a few bruises on my posterior parts and a pain in the left wrist. I became aware of the buzzing sound of onlookers, peering down at me from their vaulted height. Shakily I bent my legs and drew my feet under me. I would have arisen then if my attention had not been arrested by a redheaded specter standing demurely in front of me.

What I saw made my heart flutter. She was tall and stately, qualities that were enhanced by high heels and a high crowned coiffure and delicate but hardly notice-able makeup. That she was slender was visible even through the beige overcoat, which was belted tightly about her waist. Her soft, youthful face was deliciously framed by curlicues that had been purposely pulled from her coifed hair to provide a semitransparent view of small, red ears from which dangled circular golden earrings. Her nose was thin and aristocratic, her lips fine and drawn. Her chin ended sharply and receded to a long and languorous neck, which was bedecked with a gold chain from which suspended a single pendant: a large, red ruby that matched her hair and eyebrows. But by far her face was dominated by dazzling green eyes, the color of purest emerald. They were large and bright – if not necessarily cheery – and stared down at me almost with disdain, and seemed to transfix me to the floor like a pair of arrows.

Helping hands slid under my armpits. Reluctantly I tore my eyes away from that piercing gaze and brushed myself off as I was lifted to my feet. My helpers were dressed in flight uniform and I took them to be members of the crew. To their credit they did their best to maintain decorum and diminish any embarrassment. Not until I vouchsafed my lack of injury and apologized for my ineptitude did they smile – graciously, not condescendingly. Then they forged ahead of the crowd of people who were having their tickets checked, and disappeared through the door that led to the plane.

When I collected my senses, I returned my attention to the redheaded vision of pulchritude. She was no longer looking at me, but instead stared rather fixedly at the floor. I followed her line of vision. She was staring down at my suitcase. It had come to a halt practically at her feet. What I saw embarrassed me to no end, for the selfsame red and white boxer shorts – now limp and blackened – were dangling for all to see.

When I looked back into her eyes she was again looking at me, this time haughtily. Her expression seemed to hardly waver, except for a barely perceptible arching of one red eyebrow. Otherwise, her demeanor was one of deadpan sincerity.

Meanwhile, I turned as red as a beet. In two strides I had covered the distance to the incriminating suitcase. But when I bent over to pick it up my camera bag slid around my neck and under the arm of the chair, so that when I stood up it caught on the arm and lifted the chair up off its feet to send it crashing backwards onto a woman's carry-on luggage. She in turn uttered a sharp reproach, lifted the chair and removed her bag, while I uttered apologies.

Fearful of again locking gazes with those blazing hazel eyes, I turned with all the nonchalance I could muster toward the check-in counter. I pulled the ticket from my pocket and said in as loud a voice as I could muster, while standing over the end of the suitcase with the exposed underwear, "Flight 287, please, if I am not too late."

The girl at the counter was prim and proper and acted as if nothing unusual had occurred. "Oh, you have plenty of time, sir. Your flight won't be leaving for at least twenty minutes – until they de-ice the wings and get the runway cleared."

I yielded my suitcase, but explained that I would carry on my briefcase and camera bag. She stamped my ticket, gave me a carbon copy and a receipt for my suitcase, and said pleasantly, "You stay on the plane straight through to Hawaii, sir, where you transfer to Qantas Flight 860. The layover is supposed to be six hours, but because of the delayed departure you will actually only have a four hour wait. Your baggage will be transferred automatically. Enjoy your flight."

"Thank you," I said, stepping back away from the counter. I turned to face the place where the redhead had been standing, but she was nowhere to be seen. I didn't know whether to be relieved or disappointed. I shrugged off the feeling and I followed the last of the people who were entering the open doorway to the snowy outdoors.

I began to have an uncomfortable feeling in the seat of my pants. As my ticket was being reviewed by a clerk at the door, I slid my hands around my back. Apparently, when I fell, my coat slipped up and my pants got soaked by the wet floor. It was just another discomfort that I would have to endure. At least my long overcoat sufficiently covered the damaged area so I did not also have to suffer ignominy as well.

As I boarded the plane there were two familiar faces there to greet me. They were the two uniformed men who had lifted me up off the floor, only now I recognized them as the pilot and copilot. I flashed them a quick smile that they knowingly returned.

Halfway down the aisle I found my seat. It was next to the aisle so I did not have to disturb anyone to get into it. I stowed my briefcase and camera bag in the overhead compartment, and sat uncomfortably in my seat. The wet pants were going to be a problem. Yet, somehow I felt secure, like a baseball player who had just slid into home base after hitting a fly ball into center field. I felt that, despite the trials and tribulations of my getting here, now that I had boarded the plane, all the pressure was off me and I had nothing to do from now on other than to relax and enjoy the flight. It didn't even bother me when another forty minutes passed before we took off through clearing skies.

Interlude 1

In a one-room school there was a tendency for younger students to absorb some of the subject matter that was being taught to upper classmates. This was just as true in central Australia as it was in any other rural area of the world.

The use of the word "rural" was a stretch of the imagination. If central Australia was rural, then New York City was a backwater town. Central Australia did not possess a dense enough population to be granted the crowded status of "rural."

Outsiders tended to confuse "rural" with "primitive." Although it was true that there were conclaves of Aborigines who lived far from the maddening crowd, the typical Aborigine was anything but autochthonous. Most Abos, as they were called in an informal and nonpejorative manner, spoke excellent English courtesy of the progressive British school system, complete with what Americans arrogantly called an accent. While Abos may have associated themselves with their inherited native culture, the majority of parents did not wish to see their children educationally crippled and unable to get along in a world that was growing smaller day by day. They embraced scholastic values while harboring few provincial attitudes. The Abos represented an odd mixture of sophisticated people living in primordial conditions, doing their darnedest to bridge the finest attributes of both cultures.

Physically the Abos were an ethnic group; spiritually they were part of the past; in outlook they were reformist and forward-looking. In attitude they were a far cry from primitive.

Part of the so-called civilizing process was the bestowal of Anglican given names, juxtaposed with the retention of Aboriginal surnames. Shane Tjapaltjarri and Albert Napangardi displayed an aptitude and scientific curiosity that were unusually advanced for their age, no matter what their cultural background. Yet they were still only boys at heart. They often expressed their boyishness not by playing pranks, for they were far too serious-minded for high jinks and other mischievous juvenile activity, but by creating experiments that mimicked those of their teacher, Professor Wayward.

After the professor delivered a biology lesson concerning molds to the graduating students (all three of them), in hopes of inciting some scientific interest, Shane and Albert took it upon themselves to conduct experiments that the professor was hoping the older students would conduct.

Except for one cabinet that held toxic chemicals, there were no locked doors in the schoolhouse. Indeed, there were no locked doors in any Aboriginal community. The professor's lab was always open. After school hours, the two friends and close relatives let themselves into the lab. They brought with them a handful of medicinal herbs whose potency lay in their ability to promote healing. They carried these herbs in a cupped chunk of metal that they had found in the bush. The inner surface of the metal was coated thinly with slime.

They moistened a slice of bread and placed it in a jar. They squeezed the juice out of several herbs, the same as they would do if they were treating a deep cut or scratch, and let the liquid mix with the water in the jar. They added a "starter" mold from one of the professor's clandestine experiments. They placed a lid loosely on the jar. Then they secreted the jar and its unholy concoction in the darkest spot they could find in the back of a cabinet. They placed the chunk of metal on the same shelf.

They closed the cabinet door and left the schoolhouse.

Chapter 4

The sky was clear but turbulent. Bucking fierce headwinds, the jet shook and rattled like an amusement park ride. I drew my seatbelt tighter and eliminated all thoughts of perusing the materials in the folders. Instead I popped a Dramamine pill for motion sickness. This made me so drowsy that I could hardly keep my eyes open. I didn't fight it. I slumped in my seat, let my eyelids drag themselves over my eyes, and dreamed salacious fantasies of sharing a roller coaster car with the redhead in first class.

I woke up over Arizona in time to catch a glimpse of the Grand Canyon at sunset. The flying waitress was handing out drinks and snacks (I had missed the TV dinner), so I wet my whistle and filled my grumbling belly with salt-covered pretzels. The carbohydrate snack helped me to regain some alertness if not composure. I stared glumly out the window at the tranquility on the ground.

The pretzels settled my stomach enough to renew my energy level to minimum thrust. I opened the folder that contained my itinerary. Counting layovers, my flight to Darwin was going to take a day and a half. I closed the folder and groaned. My muscles were already cramped from spending so many hours slouched in my seat. At that I was lucky, for no one occupied the seat next to me, so I could twist my body and stretch my legs under the seat that was diagonally forward of mine.

I felt much better about my thirty-hour flight after reading an airline brochure that I found in the seatback pouch. The first settlers that colonized Australia traveled for nine *months* in the cramped hold of a sailing vessel. Each passenger occupied a space below deck that measured five feet by eighteen inches: barely enough room for a short person with narrow shoulders. The ship was unheated and without air conditioning.

People shivered from cold at the beginning and end of the voyage, and sweated interminably as they sailed through the tropics and crossed the equator.

Many emigrants never became immigrants because they died along the way: of scurvy, heat stroke, dehydration, disease, or all of the above. My sore muscles felt much better by comparison. I stretched my legs on the way to the lavatory. It was a good thing that I did, for we landed in LA so far behind schedule that there was no time to deplane to take a walk around the terminal.

A dozen people remained in their seats as the crowded aisle slowly cleared of passengers who had either reached their final destination or were transferring to other flights. I saw the back of the redhead in first class. She was going to Honolulu. As soon as the last passenger left the plane, a slew of rowdy and gaily attired vacationers rushed and crushed aboard for a trip to the nation's most recent admission to statehood. Hawaii had become the fiftieth State only five years earlier. I was forced to retake my seat before the spring in my muscles was unsprung.

As the ground crew refueled the plane, baggage handlers took turns at seeing who could destroy the most luggage. One fat fellow wearing dungarees and a red T-shirt was winning the contest hands down. Suitcases clattered to the tarmac in riotous abandon when one cart was turned too sharply and the delicately balanced load overturned. Inbound luggage received treatment that was equally as rough.

The woman who sat next to me should have bought two tickets. She spread across half of my seat in addition to her own. She was broad across the shoulders as well. Worse than that, she was talkative, and I had to listen to her inane chatter for the next three thousand miles. I opined that the settlers in the sailing vessel might not have had it as bad as I imagined. Oh for a quiet seat with legroom in first class next to a lean redhead.

Because we were going to fly over the ocean, the

stewardess issued special safety instructions in the event of what she called a "water landing." I thought to myself dryly, there's no such thing as a water "landing." When a commercial jet sets down on the water, it's crashing, not landing. Amphibious planes can "land" on the water, and then only by stretching the definition of "landing" to "watering."

This time I accepted the TV dinner because I was famished. The clamor from the rabble-rousers was louder than Times Square on New Year's Eve. They quieted down around midnight. The most difficult part of the flight was getting out of the way when the fat woman needed to use the lavatory. Her gross obesity made it difficult for her to stand and move into the aisle, where she barely fit.

The first of the Sandwich Islands to come into view was Hawaii: easternmost and southernmost of the eight major islands, and larger than all the other islands put together. At dawn's early light we flew low enough to see the volcanic plume of Mauna Loa, which was displaying mild activity. We then flew northwest over some of the other islands until Oahu's pompadour, better known as Diamond Head, cast its giant silhouette into the air. I glimpsed Pearl Harbor and Hickam Air Force Base as we circled for landing at John Rodgers International Airport.

The rowdy tourist group applauded rowdily when the tires screeched across the tarmac. They were also the first to get out of their seats despite the stewardess's warning over the public address system. I hunkered in my corner, noting wryly that the flight wasn't over till the fat lady sang. She didn't sing; she chattered like a wild turkey. I never understood why people were in such a hurry to depart, when even the last person off the plane would beat his beaten luggage to the terminal. I waited patiently until the plane was nearly empty before standing and hefting my carry-on bags. Although I was switching airlines, my suitcase was checked all the way through to Darwin.

I barely had enough time before departure to make

a phone call. I purchased a bunch of quarters from the ticket counter, then made my way to an empty phone booth. The airport noise was reduced somewhat when I closed the folding doors. I dialed zero for an operator.

After giving her Julie's number, I dropped the requisite number of quarters for a three-minute conversation in the coin slot. The quarters tinkled loudly in the collection box.

Julie answered drowsily. "Hello."

I injected cheer into my voice. "Hi, Hon. Sorry I didn't call last night but I left in a hurry."

"Left? You just got back. Where are you?"

"I'm in Honolulu - "

"*Honolulu*? What are you doing in Honolulu?"

"I'm on my way to Australia. A rush assignment - "

Julie never quibbled. She got right to the point. "How long will you be gone *this* time?"

"It's a two-day flight each way. At least a week or - "

"And what about the opera?"

"Well - "

"I thought so. You're going to stand me up *again*."

"Now, that's not fair. The last time - "

"Was the last time. Don't bother calling again."

"But - "

The call was terminated by a bad word and a loud clink. And I still had a minute left.

Despite my funk, I was delighted to find the redhead at the Qantas gate for Darwin. I sidled up alongside of her, then stepped forward into her line of sight. As soon as we made eye contact, I said, "I'm surprised that we didn't switch to Qantas in LA."

She glared as if I had said that her legs were ugly.

After a response that was cold enough to freeze an Eskimo pie, I didn't offer to shake hands. But neither did I wilt under her sour expression. I had learned from conducting scores of interviews that many people wore a seemingly grim face until I said something cute that made their lips part. A warm person often lives under an apparently hardened exterior.

"I take it you're going to wallaby land."

She kept glaring. The temperature dropped until it was cold enough to freeze the nuts off an iron bridge.

I didn't take it personally. Well, not too personally. "Hi, I'm a reporter for – "

"I know who you are. You're the fellow who is trying to take my job away from me." The resentment in her voice was obvious and pronounced.

"Huhn?" Not much of a comeback, I admit. My mouth moved as I fumbled for words, but I couldn't find an appropriate rejoinder. I was struck quipless.

Then she did a complete turnaround. "You know, if you weren't so handsome I could really hate you."

I could feel the heat rising in my cheeks. I was caught between a stammer and a blush. She had succeeded in putting me on the defensive with only a couple of dozen words. Should I apologize or ask for help?

"You know, you almost got the jump on us. But my boss found out about the whole deal – about you leaving yesterday so your magazine could get the story first. So he booked passage for me on the same plane. We weren't going to do this piece for another month. But don't think that your magazine is any better than mine. All right, we don't do as many science articles, but we do a better job. You're the big bully type. I can tell."

As you can see, she was a fast talker. I had not even had a chance to defend myself. In fact, I did not know what to defend myself against.

"Look here, Miss – Miss . . . I didn't catch your name."

Bitterly she said, "The same old approach. My name is Arlene Hawkins, as if you didn't know."

"Timothy J. Baker."

"Cut the formalities. I'm Arlene, you're Tim."

She's that kind of a girl – er, woman. She's straightforward, blunt even, and the sweetest, prettiest girl – er, woman – you ever saw. Her long red hair was tied up in a knot, more for time-saving than for beauty. She's beautiful anyway you look at her, and I looked at her many different ways. I could easily see how she got a job as a reporter. We reporters have to be forceful, you know, and she was definitely the forceful type. She was by no means elegant, and to be accused of it would probably have been an insult to her. Instead, she was cute. You know the kind of girl I mean – just plain cute. I would say cute as a button, as the saying goes, but I never quite understood how the comparison applied. A lot of buttons are not very cute.

"All right – A-A-Arlene." I don't know why, but I stuttered. I wanted to engage her in conversation so I used the trite line: "Since we're going to be working together – on the same topic, I mean – we should get to know each other better."

"Whatever you want to know I'm not going to tell you, so don't bother asking."

"S-s-sure," I stuttered again.

She looked away for a moment and then turned around apologetically. "I'm so sorry. I suppose I put you in a very embarrassing situation."

"Awkward, let's say."

"It's a bad habit of mine. I feel that everyone is against me so I put them into – " She paused. " – awkward situations. Let's start all over again. We're competitors so we can't be friends, but what did you want to know?"

I grinned, feeling slightly more comfortable now that she had changed her attitude toward me. We chatted straight through until boarding time. I learned everything about her – just about. And she learned everything about me – just about.

If you haven't guessed already, she was a reporter for the *Monday Morning Post*. She was traveling to Australia to interview Professor Warren Adolphus Wayward. Yet her purpose was different from mine. Whereas I was assigned to focus on his scientific achievements and his experiments with plant growth by means of irradiation, she was tasked with delving into his relationship with the Aborigines for her magazine's family oriented readers.

"So we're not really competitors," I suggested. "Our articles have different angles. Mine is biological; yours is sociological."

"Well, I guess if you put it that way – "

"Each piece complements the other, and both of them compliment the man."

She tilted her head so that a few loose curls bounced off the side of her perky face. "That's a very effective use of homonyms."

"Does that mean that we're friends?"

"Not necessarily. But maybe we don't have to be enemies. You work your angle and I'll work mine – "

"And never the twain shall meet."

"Just as long as you keep it that way."

"Cross my hart and hope to dye my hair black."

I think she missed the pun about the male deer. "Brown hair becomes you. Brown eyes, too."

After this repartee I considered asking her to go to dinner with me. But they announced our flight so we ate on the plane. She had first-class steak; I had economy-class chicken.

Over the Pacific Ocean there was nothing to see but water, and that not even well, since I sat next to the aisle. Anyway, I slept during most of the flight, and managed to avoid getting airsick. I was awakened by the announcement of our descent to Darwin.

The last hundred miles was pure terror. We hit downdrafts so powerful that had my seatbelt not fastened me to the seat, I would have been plastered against the ceiling. As it was, my body consistently went down with the plane while my partially digested meal did its best to maintain altitude: the food in my stomach remained stationary while my gullet soared and dived for cover. This action and equal but opposite reaction forced the victuals up my throat and out my mouth. I passed the last half hour of the flight with my lips glued to the opening of a barf bag.

In *Around the World in Eighty Days*, Jules Verne never once mentioned mal de mer.

I was pale as a ghost by the time we landed. And that was when my troubles really began in earnest.

Chapter 5

I caught up to Arlene at baggage claim. This time she looked at me with a grin instead of a grimace. And a coquettish grin at that. That took my mind off Julie.

"Enjoy your flight?" she asked.

"Not hardly." I didn't go into the nauseating details. "I could have done without the potholes and frost heaves."

"The hazards of flying, I suppose." She scooped up her suitcase as it passed on the carousel. She pierced me with a toothy smile. "You'll have to excuse me for being short, Tim, but my editor has made arrangements for me that don't include company. As much as I'd like to have you along . . . "

"Oh, oh, of course," I stammered, flustered by her abrupt announcement of departure. "Uh, *my* editor has also made arrangements for private transportation." I shrugged and raised my eyebrows. "I've got only one ticket for Alice Springs."

She held out her hand. "Then we will meet again soon." She said it in a tone that sounded like, "Let the best one win," as if we were in a race. I had a feeling that she had something up her sleeve besides a pretty arm that connected her fingers to her shoulder.

Distractedly I shook her outstretched hand. "Uh, sure thing." Then I added, "But there's only one bus to Alice Springs."

"I'm not going by bus." She gripped her grip. "Toodeloo."

I would have been angry if she hadn't been so pleasant about it. Instead, I just stood there like a toadstool, wondering why she was keeping me in the dark and feeding me fertilizer. Perhaps she still viewed me subconsciously as a rival.

Just then, a glimpse of my suitcase brought me out of my reverie. I grabbed for it and missed. My shin bumped the edge of the carousel. I lost my balance and toppled forward onto the conveyor belt. I was borne along helplessly with the luggage. A Good Samaritan reached out to help, but his hand slid down the leg of my trousers, and he succeeded only in pulling off my shoe.

My shirtsleeve got caught on the head of a protruding bolt on the inside of the carousel. The material quickly ripped free. I tried to roll over onto my knees, but was hampered by surrounding baggage. Then a tanned fist grabbed the front of my shirt. One quick yank and I found myself lying in a heap on the baggage room floor. I looked up into the stern face of a uniformed airport security guard.

"No ridin' the belts, Mate." The accent was authentic Australian. "It ain't allowed."

I scrambled to my feet, hoping that my face didn't look as red as the heat of embarrassment made me feel. I knew that it was useless to make excuses, as he must have seen me fall, so I didn't try. "I was trying to be a kid again."

He nodded once. "Right." He handed my suitcase to me. "And tuck in yer drawers. It ain't proper."

I looked down at my waist in fear, then realized that he was referring to the shorts that were sticking out of the side of my suitcase. "Just airing them out," I quipped.

I took my suitcase. The guard took his departure.

I ignored the sneers and snickers of gawking passengers. I was thankful at least that Arlene was not around to see my latest klutzy escapade. I examined my torn shirt; it was beyond repair. A thin man wearing a thinner suit handed me my shoe.

I found a quiet place to sit in order to don my shoe. While I was sitting, I examined my travel folder. From this point I was on my own. A slip of paper gave the address of the bus terminal. I exchanged U.S. dollars for Australian money, then stepped out of the concourse into the blazing Australian sun. The sky was so clear that I could see clean across the universe. A few puffy clouds, like white balls of cotton, drifted slowly across the cerulean firmament low down on the horizon. The air smelled fresh and crisp, in stark contrast to the fetid odor of bunched humanity and unburned hydrocarbons that suffused New York City.

And I wasn't even in the outback!

Already I felt better about this assignment, despite my contretemps with Julie. This was air that I could breathe . . . and in which I could run a few miles as training for the upcoming race. I inhaled so deeply that I nearly passed out from hyperventilation.

I was surrounded by a polyglot of accents and foreign languages. Locals mingled with tourists, each distinguishable more by their clothing than by their speech pattern. The gay and colorful pants and shirts of the visitors and vacationers contrasted sharply with the practical browns, grays, and blacks of the Aussies, who also wore sturdy boots or work shoes, as well as broad-brimmed hats as protection from solar glare.

I hailed a cab for the bus terminal. The driver was talkative and friendly. He informed me that there was only one bus a day for Alice Springs, and that it did not depart until late afternoon. I decided to use the extra time to shop for an outback outfit before I went walkabout in the wilderness.

The driver took me to an outfitter, and helped me pick a couple of long-sleeved cotton shirts (no one wore short sleeves in the outback) and heavy-duty trousers. He suggested cowboy boots, but said that rugged shoes with thick soles and uppers would suffice. I opted for shoes. He insisted upon a hat: not the soft felt variety with the left brim pinned up like those affected by soldiers in the war, but one made of tanned leather with an all-around brim. I went for it.

I wore one set and packed the other. I threw my damaged dilettante clothes in the trash.

The driver would not accept a tip when he dropped me off at the bus terminal. "Enjoy your stay, Mate. You'll like it here."

"I like it already, and I've only been here an hour."

He took my suitcase out of the trunk with an ever-present smile. "They always do. A lot of folks never leave."

He stuck out his hand. I shook it.

I purchased my ticket and sat down to wait in the adjacent cafeteria. I found that I was almost too exhausted to eat. Having crossed the International Date Line, the local time was a day later than my biological clock was telling me. I was so drowsy that I barely heard the departure announcement. I boarded the bus, found an unoccupied row, and quickly lapsed into a deep sleep. I was content to be on the ground instead of pitching and wallowing in the air.

Alice Springs was smack dab in the middle of the subcontinent. It was located more than eight hundred miles south of Darwin. I slept soundly throughout the night, missing the whistle stops that offered food and opportunities to stretch the legs. When I opened my eyes and glanced at my wristwatch, I had no idea what time it was. My watch was still set on New York time. Judging by the height of the sun it must have been late morning. A fellow passenger gave me the correct local time. I pulled out the stem and reset my watch.

Towns and people were few and far between. Although Australia's landmass was 85% of the size of the continental United States, the population was less than that of New York State alone.

I remembered something that I had read about cordiality in relation to population density. When people were packed together like rain-soaked cats in a small canvas sack, they tended to fight viciously in a bid for more personal space; hostility bred faster than blood-glutted fleas. When people were spread across the land out of sight of the nearest neighbor, they were lonely and therefore happy to meet another human being. Ergo the difference between neurotic city dwellers and the friendly inhabitants of rural communities. It made sense to me, especially now that I was seeing the evidence firsthand.

We stopped at a way station whose name I never learned. Both the entry sign and the departure sign were nailed to the same post. The town consisted of a few outbuildings and a gasoline station with a snack counter. Seems as if most of the passengers traveled with their own food. I bought some victuals and a gallon jug of water. The bus refueled with petrol, as the locals called gasoline, and a fresh driver took the wheel from the previous one. We were on our way after less than ten minutes.

I had a hazy notion that the central Australian outback consisted largely of desert. This notion was confirmed by the long ride through territory that was nothing but sand and scrub vegetation, although stands of trees occasionally punctuated the monotony.

The heat was now unbearable. The bus was not air-conditioned. Open windows created cross-flow ventilation that did little to dispel the heat but much to assault the ears with high-speed noise. Dust permeated every pore.

By the time we reached Alice Springs late in the afternoon, my sweat-soaked clothes and exposed skin were coated with dust. Now I knew why brown was the predominant color for outerwear: it disguised the dust and presented an appearance of cleanliness.

Alice Springs was larger than I expected it to be. Instead of a two-bit town that was hardly worth noting on a map, it was a thriving if small municipality that consisted almost entirely of suburbs. The population boasted more than ten thousand people, to say nothing of cats, dogs, and kangaroos. Once the outback was no longer in sight, the only things that distinguished it from Smalltown U.S.A. was the mode of dress and the accent of the dressers. The MacDonnell Ranges straddled the town, extending east and west for hundreds of miles.

Here I was left hanging. Gadfly's arrangements ended with my arrival in Alice Springs, with the curt suggestion that I "obtain local transportation" for the remainder of the journey to Professor Wayward's isolated schoolhouse. When I questioned the station master, I learned that the professor was rather well known by the inhabitants. He told me the general location of the schoolhouse, and suggested three ways of getting there.

First: the owner of a sight-seeing plane would gladly fly me over the outback for an exorbitant fee, but as there were no suitable landing strips in the vicinity of the schoolhouse, he probably would not be able to land anywhere nearby. Second: a supply truck that ran on an approximate weekly schedule would be venturing to the schoolhouse in five days time to deliver food and supplies. Third: I could hire a native guide to lead a safari through the outback to the village.

That didn't leave me much of a choice, but I ruled things like this. The first was definitely out because I had no stomach for flying in jetliners much less small private planes; and besides, if the pilot managed to find a place to land, I would have to walk the rest of the way, and it left me stranded out there with no certain way to return. The supply truck was not leaving right away, but was continuing on to other villages for trade; it did not follow a strict schedule, but returned only after all its supplies had been traded for native goods.

This gave me pause to wonder about Arlene's method of transport. Hertz did not rent Land Rovers in Darwin – or anywhere else on the subcontinent.

My last resort was a safari which I would have to keep on the payroll until my return to Alice Springs. So it looked as if I had to hike it.

By this time, most establishments in Alice Springs were closed. I found inexpensive lodging in a two-story hotel that catered to the tourist trade, such as it was, and an equally inexpensive diner. I was famished; the meal was excellent.

The hotel was called the Wallaby Roost. It was aptly named, because the rooms were as small and as dark as a wallaby's abdominal pouch. I awoke refreshed after a good night's sleep under thick blankets and a fluffy quilt. I asked at the front desk if the night's were always that cold.

The kindly matron said, "This is a desert. Heat dissipates rapidly when there ain't no vegetation to keep it on the ground."

Fortunately I had sweaters in my suitcase from my trip to Austria and winter in New York. I donned the heaviest one before venturing outside for breakfast. The stark blue wool clashed harshly with the rest of my attire, but I didn't care. I was warm.

The coffee was so strong and bitter that it reminded me of reheated espresso. I pushed the cup aside and ordered orange juice to drink with my steak and eggs. Afterward I walked to the address of the safari outfitter. The crisp air was invigorating.

The person who greeted me was an Aborigine bushman who wore a standard outback outfit of faded tan, and who spoke perfect English; or rather, perfect Australian. He overworked the local courtesy title, pronouncing it "mite" instead of "mate." In fact, he pronounced all of his long a's as i's. It was a cute touch that I found infectiously appealing. I was beginning to love this country despite my forebodings about coming here in the first place.

"The nime is Charles Yarramunua, Mite. And what can I do for your lordship today?"

"I – I – I'd like to organize a safari."

"I can do that, Mite. Have you already procured your hunting license?"

"I – I'm not here to hunt. I'm a photojournalist. My name is Timothy J. Baker, from New York."

"Ah, shooting the wildlife with a Canon instead of a Remington."

I thought about the Canonflex RP in my gadget bag, my favorite single-lens reflex camera, with a host of interchangeable lenses for all occasions. "Yes, how did you know?"

"Common knowledge. Nikons don't stand up well in the dust."

I nodded as if I was aware of that fact all along.

"My advice is to keep your camera wrapped in plastic when not in use. Lenses too."

"I'll do that."

Charles placed his hands on the counter top. "What

kind of animals would you like to shoot?"

"Well, I'm not really here to shoot animals. I'm here to shoot – uh, that is, to photograph and interview a Dr. Wayward."

"Ole Doc Warren. I know him well."

"You do?"

"Absolutely, Mite. He's doing great things with my people. I furnish his supplies on a regular bisis, and provide transportation whenever he's needed in – well, you know, here in town."

"Oh, he comes here regularly?"

"Not regular. Only when he's called."

I had the impression that Charles was hinting at something unsaid. I let it pass. "So, can you get me out to his place? The schoolhouse?"

"That I can do. It is located about forty-five kilometers southeast of here."

I couldn't think in the metric system but I could convert meters and kilometers to feet and miles in my head, although not easily and only approximately. For the convenience of my readers, I will do the conversions for the remainder of this narrative, but keep in mind that even though I quote Australians in U.S. customary measurements, they spoke in metric units.

I groaned at the prospect of bushwhacking thirty miles through Australian outback, baking under a hot sun during the day, camping at night under a cold firmament, and lugging a heavy suitcase. I'd have to hire Sherpas, or whatever they called porters in this neck of the woods. I had visions of Africa and Henry Stanley on his quest to find Dr. Livingston among the natives. Suddenly I wasn't liking Australia as much any more.

"How – how long will it take to get there?"

Charles closed one eye and cocked his head to the left, calculating mentally while staring sightlessly at the ceiling. He took a long time in his figuring. I could see the days mounting. I was almost afraid of the answer.

Finally he breathed, "About four hours."

"Four hours!"

"Yes, you see, the bridge is out over Noname Creek

in Damifino Canyon, so we have to drive about fifteen miles out of the way to cross the ford in the Noplace Valley, then meander along the left bank until we intersect the road. It adds an hour or more to the trip. It's slow going when you're off the road."

I could have hugged Charles. I breathed such a deep sigh of relief that my exhalation blew loose sheets of paper off the counter. "Wow. I thought we were going to have to *walk* there."

"Nobody walks, Mr. Biker. Except the wild Abos. If the terrain is too rugged for Land Rovers, we ride horses. Now, if you want to go on horseback, I can arrange it for you."

"No! No. A Land Rover is fine. Perfect, in fact. It's just that I thought a safari was, well, porters with baggage on their heads, and all that."

Charles laughed raucously, showing a mouthful of fine teeth so dazzling white that their brilliance was blinding. "This isn't the Dark Continent, Mr. Biker. We're civilized here." Then, as an afterthought, "For the most part."

I didn't realize how tense I was until I felt my body sag and my taut muscles relax. I made a show of wiping nonexistent sweat off my forehead. "You just made my day, Charles."

Charles nodded knowingly. "Perhaps not, when I tell you the cost."

"How much?"

He quoted a daily rate in Australian money which, when I converted the amount to U.S. dollars, seemed quite reasonable. Besides, I wasn't paying for it anyway. The money went on the expense account. I just hoped that the story in its final form was worth the cost of producing it.

"Sold to the man with the Land Rover."

"When did you want to go, Mr. Biker?"

I splayed my hands. "Any time."

"Give me ten minutes to pack some food and water."

"Wait. I've got to get my things from the hotel."

"We'll collect them on the wiy out of town."

Things moved swiftly with neither nonsense nor formality. Charles tossed some tins of food, a loaf of bread, and three full canteens into a canvas knapsack, and placed it between the bucket seats. He filled a pair of five-gallon jugs with fresh water, which he secured to the two five-gallon jerry cans of gasoline that lived permanently in the back of the Land Rover, along with camping gear, a huge tool kit, and several boxes of spare parts. We were ready to go.

Charles idled the Land Rover in front of the hotel while I grabbed by belongings and checked out of my room. He topped off the gas tank at a petrol station, and took time to check the oil level in the engine and the air pressure in the tires. The paved road ended and the outback began less than a mile outside of town. At moderate speed we bounced along a well-worn dirt trail that was smooth and hardly rutted.

The outback invigorated me. I felt as if I were commencing a great adventure into unknown territory. I had never been in the wilderness before, so it was a new and exciting experience for me. The obligations of civilization fell off my back like leaves from a tree in autumn. At the moment, the assignment seemed less important to me than the thrill of sudden freedom from the shackles of responsibility.

I thought about my boss's rush to send me on this assignment. Granted that Gadfly was a chronic go-getter, but something about this assignment didn't jive: his rambling about the *Monday Morning Post*, about the importance of this article on Professor Wayward, about occurrences "higher up." Then there was the statistically improbable coincidence of meeting my competition on the plane.

I wondered, exactly who was trying to get the scoop on whom? A hammering at the back of my mind warned me that something fishy was going on back home. Otherwise, how did both magazines get the impression that the other was trying to beat it out of a story? It didn't make sense. And why didn't Arlene travel on the same – and only – bus to Alice Springs that I traveled on?

My thoughts were disrupted by a group of kangaroos that jumped out of the bushes in front of us. Charles slammed on the brakes as I fumbled for my camera. Quickly I switched lenses and mounted the telephoto. I snapped a few shots for local color. Mostly I got tails and backbones as the kangaroos bounded away like giant brown bunnies on pogo sticks.

"That was close," I whooped. "We almost hit them."

"Not to worry. We've got a roobar."

I looked at him askance. "A what?"

He laughed, then pronounced the words slowly: "A – roo – bar."

"What's a roo bar?"

He pointed a firm black finger at the chromed tubular framework in front of the grill, surrounding the winch assembly. "That's a roo bar. It's like the cow-catchers on locomotives in your Wild West movies. It pushes the roos out of the way."

I nodded in understanding. With the excitement over, my thoughts leaped back to Arlene. "Do you get many tourists out here?"

"Fair amount. This is our slow season, because of the heat. The outback is nicer in the winter."

"Uh, anyone recently?"

"Not for a fortnight. I've been busy doing tune-ups and repairs."

"What about other outfitters."

"I'm the only gime in town. Not enough business to support more than me."

So Arlene had yet to reach Alice Springs.

"I thought I told you to wrap your camera in plastic."

I shrugged. "I didn't have time." I placed the camera back in the gadget bag. "Now it's too late."

"Ask Doc Warren. He'll give you some."

Charles proved to be a gifted guide. He was as affable as he was knowledgeable. "Before the Abos migrated to Austrilia, the land was inhabited by marsupials and monotremes. No mammals except for bats and other small rodents. In the evolution of life, marsupials and monotremes preceded the mammals. Because a marsupial's young are born before they are fully developed, the mothers carry them around in a pouch, where they suckle them. Marsupials include kangaroos, wallabies, phalangers, wombats, bandicoots, thylacines, dasyures, and banded anteaters.

"Then you have your monotremes – the only ones that are alive in the world: oviparous anteaters and duck-billed platypuses. These protomammals liy eggs like birds, but unlike birds they suckle the hatchlings.

"Birds that you don't have in the States include parrots, toucans, and flightless emus and cassowaries which, due to adaptive evolution, let their wings wither away to complete uselessness.

"At night you'll hear the baritone howls of dingoes – a kind of feral dog – but they were imported, like the rabbits."

Vegetation was extensive. It consisted of palms, tree ferns, orchids, acacias, and spectacular three-hundred-foot-high eucalyptus trees. We didn't see any eucalyptus trees, but Charles described them as giant umbrellas that enveloped the ground with wonderfully cool shade. He pointed out dozens of other plants as we passed them along the side of the trail. He was a naturalist and a natural storyteller: the kind of guide that a tourist would want to have on a safari into the hinterland.

I jotted down notes as fast as I could write. The bouncing and swaying of the Land Rover made much of my scribbling illegible, but I figured that I could transcribe my notes later when I was sitting on solid ground.

We stopped at the top of a gently rising hill with a view of Damifino Canyon, Noname Creek, and what was left of the bridge after a flash flood had torn out the planks of the center span. Most of the wooden beams remained in place, but it wasn't possible for vehicles to cross the bridge until the missing planks were replaced, the piers were rebuilt, and the abutments were strengthened.

Charles grabbed the food sack and canteens, and led me to the shade of a short tree. As we ate our lunch from this vantage point, I could see some of the vastness of Australia. The land was so broad that it looked like a plain filled with scrub brush. Isolated stands of trees marked waterholes, while snakelike groupings indicated dry creeks and streambeds that would have had water in them had this not been summertime. It reminded me of Monument Valley without monuments.

Not far in the distance, both north and south, the McDonnell Range stretched as far as the eye could see. The peaks were tall monarchs that extended from one to the next for more than one hundred fifty miles. I took some wide-angle photos of the impressive scenery.

Charles chatted endlessly. His knowledge of local flora, fauna, and minerals was nothing less than awesome. He was a walking encyclopedia whose nonstop instruction exceeded my ability to absorb the information that he imparted so effortlessly.

I never doubted his facts or his wisdom. "How do you *know* all this stuff?"

"I was born and rised here, Mr. Biker. Foaled in the outback, reared in the city. My education consisted of the best of both worlds. How about you?"

"I got the worst of one world. Born in a hospital, raised in a tenement, lived my whole life in the City. I grew up in a good neighborhood, though. And my father worked hard in building maintenance. We weren't rich but we were never broke, so I guess I can't complain." After a pause, "So, are these your people the professor is teaching?"

"Not my clan. I come from South Australia, not the Northern Territory. Down Woomera wiy, before it was a rocket test site."

"I didn't know they tested rockets in Austrilia."

"Most people don't. Middle of nowhere. If a rocket goes astray, nobody gets hurt but the plants and animals – unless a wild Abo tribe happens to be in the area. Lost one a month or so back, so they say. Word is the gyros went out. Never did find it. Probably never will. Or by the time they do, it'll be nothing but a hunk of rusted metal that even the Russians couldn't put back together. Which is the point."

I was quickly learning that the land Down Under was more glamorous and complex than American propaganda had led me to believe. For a one-time penal colony, Australia was doing all right for itself.

Charles repacked the food. "Well, Mr. Biker. Let's be on our way."

The next thirty miles was the worst of the journey. We drove fifteen miles south along the right bank of Noname Creek. The Land Rover swerved from side to side as if the helmsman was a drunken sailor. There was no trail to speak of, so Charles had to drive around shrubs and avoid the worst ruts and cavities. We dipped down into deep gullies, crossed dry sandy washes, climbed over boulders, and sometimes rolled along smooth rock or hardened clay. I wondered if we were going to have to use the spare tire that was bolted to hood – or bonnet, as they call it locally.

Occasionally we approached close enough to Damifino Canyon that I could look down and see the water. The canyon walls grew shorter as Noname Creek neared Noplace Valley. The canyon petered out to the flatness that dominated the valley.

I was apprehensive when we reached the ford. It was marked only by a post with the skull of a horned bull on top of it: a dire warning, it seemed to me, that you crossed at your own risk.

The creek had spread to a width of several hundred feet, making it look more like a river. The water no longer gushed through a narrow channel, but flowed slowly, almost imperceptibly.

Charles did not dash into the shallow water, but paused to survey the submerged terrain. He got out to lock the hubs on the front wheels. Back in the driver's seat, he engaged the four-wheel-drive by shifting the the transfer case lever. Slowly he eased out the clutch and pressed his foot against the accelerator pedal. He angled the Land Rover downstream so the water did not splash higher than the tires. The creek bottom was hard-packed, so we had no difficulties with traction, although a couple of times we hit soft spots that caused the front tires to spin fast and spew a muddy froth.

After we reached dry ground on the other side of the creek, Charles disengaged the four-wheel-drive and locked out the front hubs. We meandered north along the left bank for five miles or so, then veered eastward. Because we didn't go as the crow flies, the distance to the schoolhouse was nearly doubled.

"We'll intersect the road this way, then follow it southeast," Charles explained.

The ride was smoother once we got back on the dirt road. A couple of hours later we topped a low rise, and the schoolhouse came into sight. It sat on a mound from which the surrounding vegetation had been hewn away so as to afford more freedom from the tangle of surrounding terrain.

The schoolhouse was not an impressive sight. It looked as if it had been slapped together with boards that had been taken from previous construction. It was dusty and dank and hastily built, but prodigious layers of paint disguised the worst of the ill-fitted planks. Huts or cabins guarded three sides of the building. I could hear the thrum of a generator that was out of sight in an outbuilding.

I could see a small village about a block away. (To city dwellers, a block is one-tenth of a mile.) Aborigines were occupied in various tasks, but they all stopped what they were doing and looked at the Land Rover. Charles waved. Several Aborigines waved back.

I stepped out of the Land Rover. My clothes were coated with dust and sand. I brushed off enough to start another Sahara Desert. Charles alighted softly.

From behind me I heard a high-pitched voice. "Well, hi there. Have a nice trip?"

I turned and saw a bearded man whom I took to be Professor Wayward.

Next to him stood Arlene.

Chapter 6

Stunned speechless, I glared at the two of them for a moment while I tried to gather my wits. I was witless.

Then Professor Wayward broke the silence by saying, "You must be that other reporter, from the *New York Script*, that Miss Hawkins was telling me so much about."

Despite my anger, I feigned an expression of delight as I stepped forward and stood lightly on my feet. "Yes," I said, parroting Livingston's reply to Stanley's presumptuous question. "My name is Timothy J. Baker."

I extended my hand; the professor grasped it firmly. His close-cropped beard did not hide his warm-hearted smile. I had halfway expected him to be a doddering old scientist, bent over in the last throes of subsistence. Instead, despite thinning gray hair and the beard that he had not affected in his photograph, he appeared vibrant and vigorous. His eyes sparkled with energy. He wore a white lab smock over gray trousers and a gray shirt that was casually unbuttoned at the collar.

Looking over my shoulder, the professor said, "It is good to see you again, Charles."

"Sime here, Doc."

"Mr. Baker," said the professor, bowing slightly, and returning his attention to me. "I believe you have already met my companion, Miss Hawkins, from the *Monday Morning Post*?"

She was dressed from head to foot in khaki. Her long hair was tucked under an Aussie bush hat. Brown jump boots several sizes too large hid her tiny feet. Despite the combat ensemble, she somehow managed to maintain a semblance of femininity.

"Yes, I have."

"Good, then, I am glad to have you with us. As you can see by the looks of the schoolhouse, I need all the publicity that I can get." The professor was quite candid. "My own resources are, shall we say, somewhat limited."

I could barely keep myself from glaring at Arlene. I put my best posed smile on my face. "I really hope that I can help in that regard, Professor."

"I'm sure that once the world learns that such primitive conditions exist in this modern day and age, they will back my endeavors with glee. Hopefully that will encourage some government funding as well."

"Exactly the point I will try to get across to the public." As soon as I said it, I remembered that the focus of my article was on the professor's scientific endeavors, not his humanitarian efforts toward educating the Aborigines and bettering their station in life. That was Arlene's slant.

"Excellent. Excellent. Oh, by the way, Mr. Baker, I am afraid that I was not expecting two reporters from conflicting magazines and opposite sexes. I have only one spare room which, of course, Miss Hawkins is already occupying."

"That's quite all right," I said, although it wasn't. I had arrived less than one minute ago, and already the situation was out of control.

Charles came to my rescue. "He can use my spire tent and sleeping bag."

The professor glowed. "Why, thank you, Charles. How gracious of you."

I took my eyes off Arlene long enough to grin at Charles, but I'm afraid that my grin came across more as a grimace. "Yes, thanks a lot." Then, after an icy pause, "You have spares?"

"Nobody travels the outback without a full assortment of camping gear, ritions, and plenty of water. In cise of brikedown. I still remember how to live off the land, but most tourists would die in a matter of diys."

"I'm sure I'll be very comfortable in your tent. Can I share some of your rations?"

Professor Wayward held up both hands, palms outward. "Tut, tut, Mr. Baker. I am well furnished with victuals – local as well as store-bought. You will not suffer on that account. My housekeeper is an excellent cook."

"Thank you, Professor Wayward." I stared again at Arlene. "How did you – "

The professor interrupted, "Then it is settled, Mr. Baker. Let me extend my greetings by inviting you to dinner tonight. You too, Charles, unless you plan to depart immediately."

"I'm on safari as long as Mr. Biker wants me." Addressing me, "I'll get the camping gear and pitch the tents." To the professor, "Doc, can I lock the extra petrol cans in the generator shed?"

"Of course. We don't want the locals sniffing the fumes for amusement." To me, "It induces mild euphoria and keeps them calm, but it's not good for their brain cells. Shall we say sevenish, then?"

Automatically I glanced at my wristwatch. It was already three o'clock. The long lunch, longer detour, and rough terrain had made the drive six hours instead of four.

"That will give us plenty of time to set up camp," said Charles.

"So it shall be." The professor turned and walked toward the schoolhouse.

Arlene lingered behind.

Charles commenced his safari leader chores. As soon as we were alone, I pierced Arlene with a dagger-like stare.

In her sweetest voice, Arlene singsonged, "Did you have a question, Tim?"

"You're dark right I did." The timber in my voice was falling faster than a toppling tree. "I'd like to know how you got here ahead of me."

"Oh, that. I simply took a plane and parachuted. Well, see you at supper. Toodeloo."

I wanted to toodeloo her up side the head, but I restrained myself. She was the most exasperating female I had ever met. Ever. Nonetheless, she was still easy on the eyes and oozed sex appeal.

If I had half a mind . . . but I was a man of two

minds with respect to my archrival.

I steamed as I helped Charles to pitch the tents. Actually, I was more of a hindrance than a help. But he never complained.

*   *   *

Supper that night was a delicious blend of freshly butchered meat and garden vegetables, plus delicious home-baked bread. All had been prepared by Marie: the professor's combination housemaid, handygirl, chef, waitress, and general factotum. Marie was undertall and overthin: an elderly Aborigine widow from the village who had elected to work for the professor for no wages other than the pleasure of his company. She did all the domestic chores that a housewife normally does around the home: cooking, cleaning, laundry, and – well, perhaps not everything. But one never knows, and I didn't inquire.

After setting up camp, I watched Marie bake bread in a reflector oven. This ingenious device consisted of two sheets of aluminum that were hinged to open at right angles to each other. She placed raw dough on a flat sheet of aluminum that bisected the angle. She erected the oven in front of a vertical fire so that the dough sheet was horizontal. Heat was thereby reflected from the outer sheets that were now positioned at forty-five degrees with respect to both the fire and the dough, thus baking the bread simultaneously from both above and below.

This contrivance amazed me. Charles informed me that he never traveled in the outback without a reflector oven, so he could bake bread substitutes made from local plants. I had the impression that possessing such survival gear was as common in the outback as it was for me to carry change in my pocket for vending machines and parking meters in the City.

The dinner table was made from rough-hewn beams and planks. We ate off crudely thrown ceramic ware and wielded pot-metal utensils. In my opinion, this accepted lifestyle stood only one rung above camping.

I slipped my hand into my pants pocket and

pressed the tape recorder's 'on' button.

Professor Wayward had already expressed his desire that we should not call Marie a maid or, banish the thought, a servant. If she were at any time mentioned in our articles she was to be referred to as "an altruistic woman with a heart of gold, and without whom my work would have been very much diminished."

"How about 'trusted girl Friday'?" I suggested.

The professor cogitated for a moment. "I like the sound of that," he whispered cautiously, in a crisp British tone with such precise enunciation that it left me, with my ugly New York eliding twang and soft pronunciation, breathless with envy. "Yes, that will do quite nicely."

I wanted to add "and concubine," but thought that the professor might not appreciate my warped brand of humor, especially if the sentiment struck too close to home.

Charles munched thoughtfully. "Don't call her anything that will chise her awiy. After this meal, I'd hite to have to survive on tinned meat or wild tubers mixed with kangaroo steak."

"Only one of her magnificent qualities," said the professor. "Mr. Baker, I regret that you were so late in arriving, as Miss Hawkins informed me that you landed at Darwin simultaneously. Now I shall have to repeat some of the information which I have previously imparted to her. After all, she has had a forty-eight hour head start on you."

"I regret it too, Professor. But if you don't mind, I'd like to interview you in private, instead of over the common dinner table in mixed company."

"I know when I'm not wanted," said Arlene, as she rose and started to leave.

But the professor held her back with a touch on her forearm. "Nonsense, Miss Hawkins. I wouldn't hear of your leaving. As long as we are all together, you can both profit from each other's inquiries. I shall show favoritism to no one."

Arlene reseated herself, sneering at me defiantly as she did so.

Which was just what I did *not* want. Well, I wasn't about to ask any truly important questions and let her have the advantage of me. For now, I decided to keep our confabulation in a general vein.

"For a starter, you can tell me how and why you came to this remote part of the world to educate and elevate primitive people."

"I sometimes wonder about that myself. I certainly never harbored any intentions of immigrating to Australia and courting sympathy with the local inhabitants. It was purely a matter of circumstance. Let me give you some personal background so you can comprehend my circuitous path to this recent turning point in my life.

"As a boy I lived in a disreputable neighborhood, replete with ignorance and indigence. I moved away from there when I was ten, but those formative years left a lasting impression on me. In university I studied biology because I desperately wanted to understand how an organism that was so complex could sustain the anomaly called life. I was constantly haunted by my childhood experiences of suffering, because it struck me as odd that the most intelligent creature best-suited to his environment could survive in such misery. Do you realize that all so-called inferior creatures in the animal kingdom live in better circumstances than man?

"Anyhow, I also studied physics. I was working as a biophysicist when the War broke out. I was young enough to be drafted, but the government saw fit to offer a civilian position to me if I did not mind relocating for the duration of hostilities. The army had sufficient cannon fodder but was short on scientists. I worked for several months on top-secret projects at Bletchley Park, in a strictly managerial capacity, before being reassigned to a biological weapons laboratory."

The professor pronounced "laboratory" the way Boris Karloff did in his monster movies: with four syllables instead of five, with the accent on the second syllable instead of the first, and with "tory" elided to "tree."

"Due to the nature of the work and its potential for accidental catastrophe, it was decided that certain, uh, unorthodox experiments should be conducted in utmost solitude and secrecy: away from habitation in case the, uh, experiments should go awry, as well as away from the prying eyes of the enemy. We moved our entire kit and caboodle to the sparsely populated rangeland of Maralinga, in South Australia.

"Despite the isolation of the locality, we were honor bound to protect the few dwellers who scratched their livelihoods from the land: ranchers who grazed sheep and cattle on the plains, and Aborigines whose ancestors have occupied the Maralinga area for millennia. These people were displaced, leaving unpopulated a tract of land that was greater in size than the entire island of England.

"The ranchers stayed away because they understood the necessity for relocation, and because they accepted the authority of their government. They built new homesteads in designated areas, and continued to graze their sheep and cattle as they had before the war. The Aborigines, however, were not as cooperative. I do not mean to imply that they consciously objected to the new regime. Not at all. No, they simply did not – could not – comprehend the perils that were inherent in our tests.

"Many tribes were nomadic. They roamed the land the way their forefathers had done. We kept removing them from our testing areas when they strayed too close to danger, but they kept wandering back. During these frequent encounters I treated the people for common diseases that I was competent enough to recognize, and tended to their wounds – although I daresay that their poultices made from local herbs generally proved more efficacious than the modern medicines that were at my disposal.

"I studied their language and learned the local dialects. Once I could communicate with them, I commenced to instruct them on the vicissitudes of life in the modern world with which they were colliding. Charles, here, was one of my aptest pupils."

"Right on," cheered Charles.

A quirky smile appeared on the professor's face. "After the war, the character of the experiments turned in other directions. In a joint venture, the Americans were invited to share what came to be a testing range. I could have returned home at that point in time, but chose the option to retain my position – ostensibly to further my studies of affected wildlife, but in reality to continue my association with the Aborigines. An outpost was built to house the large influx of workers, scientists, and managerial staff – both civilian and military – and to provide support facilities for the various, uh, activities in which the governments were engaged. The outpost quickly grew into a town that was appropriately named Woomera: the Aboriginal word for a hooked wooden stick that is used to hurl a spear or dart.

"I worked there in various capacities and for various experimental services until the facility was officially closed last year. Although I have been retained as a part-time biohazard consultant for cleanup operations, I now have more time to devote to helping these primitive folk – not to civilize them, oh dear me, no, for they can survive quite well without the benefits, and deficits, of what we call civilization; but to teach them how to survive *despite* the incursion and influence of modern civilization.

"I want to nudge these innocent people onto the proper path of social development before they are corrupted by the greed and selfishness that characterizes human cultures in the so-called civilized world. I want them to keep their simplicity and humaneness."

When the professor paused for a moment of quiet contemplation, I interjected, "Fascinating. So you are the right man in the right place at the right time to save the Aborigines – " I glanced sideways at Charles. " – from the evils of exploitation."

I detected a faint click in my pocket. I excused myself to go to the rest room – or as the professor called it when he gave me directions, the WC (water closet). In the outhouse, I inserted a blank tape in the recorder, and placed the full tape in my other pants pocket. I started speaking as soon as I returned to my seat at the table.

"Now, don't take this personally, Professor Wayward, but a couple of years ago I interviewed an anthropologist who had discovered a primitive tribe deep in the jungles of South America. Brazil. This Neolithic tribe consisted of only a couple of dozen individuals who were completely isolated from all extraneous human interference. Had been for generations, from what the anthropologist could determine. Their lack of human contact gave him with the perfect opportunity to observe primal communal activities that were totally uncontaminated by interaction with other tribes.

"He figured on using this tribal family as the subject of his Ph.D thesis. He observed their activities without interacting with them. He observed how they lived, how they ate, how they slept, how they gathered food – everything. They accepted his presence as if he were some kind of – pet, or wild monkey, that had taken residence nearby. Because he was Caucasian, they didn't appear to understand that he was as human as they were. They tolerated him because they were so docile.

"Meanwhile, during months of observation, he saw them get sick, he saw them get hurt, he saw them suffer – from festering wounds and gangrenous infections that he could easily have cured with the medicine in his first aid kit. Instead of helping them or easing their pain, he watched them die slow and painful deaths. All so he could write a damn report and be respected by his peers. He claimed that he was 'preserving' their social structure by not making physical contact, but in fact he was sacrificing them for his own personal motive. What arrogance!"

Arlene chimed in, " 'The Savages of Academia' was the title of the article."

"You read it?" I was astonished to say the least.

"I admired the stance you took."

I wish my editor had admired my stance as much as Arlene had. Gadfly published a bowdlerized version that cut most of the meat out of the piece (although he used the word "vituperation" instead of "meat.")

"Yes, well, thanks. I think." I turned back to the professor. "As I said, don't take it personally."

"I do not," the professor said lightly. "Our situations are not analogous. I am trying to prepare the wild Aborigines for the inevitable cultural collision, not to cripple them the way the Quebecois cripple their children, by not permitting them to learn English, ensuring that they can never achieve equality in an English-speaking world. On the contrary, I want the Aborigines to be able to reap the benefits of education in a global society, while retaining the naivety of their value system.

"Right on," cheered Charles.

I stared at Charles wordlessly. If he was a standard product of the professor's provincial schooling, then I had to admit that he was making remarkable progress in his altruistic ambitions.

Professor Wayward shrugged nonchalantly.

"Yes, well, I guess I got a little off course. Your relations with the Aborigines are more in Miss Hawkins' realm than mine. My piece is slanted toward your scientific endeavors: what you've done, what you're doing, what you hope to do in the future."

"I understand. I am afraid that the work that I have done for the military is classified. The Official Secrets Act prevents me from discussing my work, or even from admitting that I was involved in such work."

"Spoken like a true Ciceronian." I was gratified to see the professor's face brighten at my evident understanding of his nonadmission.

Cicero was a Roman orator who was known for addressing the Senate by commencing with, "I will refrain from mentioning . . . " and then stating precisely what it was that he was not going to mention.

"You surprise me," said the professor. "Most Americans would not have appreciated my roundabout candor."

"Not the benefit of a classical education, I assure you, but the result of four years of high school Latin," I acknowledged. "Root words led me to a greater comprehension of English." Then, "What you've tackled is greater than the labors of Hercules."

"Perhaps, but I do take credit for some small measure of success."

"Right on."

This time I broke down and snickered. "Okay, okay. Enough bantering among the intelligentsia. I won't embarrass you by asking for classified information. What nonmilitary scientific projects are you working on now, and where do you hope they will lead?"

Arlene produced a tablet and pencil from her handbag, and commenced to take notes.

"Those are questions that I can answer. In Australia there are marvelous advancements to be made with respect to biological agents. This isolated continent holds many secrets, most of which are good, although some are indubitably bad. Many of the things that I have learned, I have kept in strictest confidence with myself, for modern man would surely misuse them for the purpose of senseless wars and fighting: venoms and poisons and neurological toxins that can kill swiftly and painfully.

"Aside from my military work, my personal specialty has been the study of rare herbs which can be found only in the outback, and generally only by Aborigines. I collect these herbs and perform analyses on them in order to explore their uses and abuses. The latter I have not divulged, but there is a whole world of herbology waiting to be discovered and put to proper applications."

The professor warmed to a subject that was obviously close to his heart. Arlene scribbled furiously on her tablet while I sat calmly and listened for inflections and nuances that I could later transcribe onto paper. In addition to recording faithfully every word that the eminent professor articulated, I had the supreme satisfaction of completely bewildering Arlene by my failure to take notes of the conversation.

The tape recorder was a lifesaver once the professor started to name the herbs from which he had extracted exotic chemical substances. These herbs all had Aboriginal names, but did not have common English equivalents. Most were unknown to Western science. Using binomial nomenclature, the professor provided the Latin words that described the genus and species. For herbs that he was the first to describe, he had created names that were based upon his scrutiny of their taxonomy.

"I have kept meticulous notes and drawings," he concluded, "in hopes of publishing a book on the vast variety of Australian herbs. My grand hope is to isolate organic chemicals that have therapeutic applications. Eh, Charles?"

"Right on. My people use extracts from many herbs as salves for wounds, burns, boils, skin afflictions, even sunburn."

"Sunburn!" I expostulated. "I thought your skin was too black to get burned by the sun."

"Generally true, but at high elevations my people sometimes suffer from ultraviolet ridiation. We also have a balm that soothes the eyes from intense UV."

"The same ointment cures conjunctivitis and similar eye inflammations," the professor added. "Ophthalmic solutions represent only the tip of the medicinal iceberg that may be found in Australian herbs." Parenthetically, he said, "On a more mundane level, herbs are also used to season food, such as the victuals that we ate this evening."

The one-sided conversation raged on. Professor Wayward expounded upon his personal thoughts and pursuits with regard to herbal essences that he extracted from transplanted herbs that he grew in a small hothouse.

Arlene and I contributed little to what was primarily a discourse. Charles inserted comments from an

Aboriginal viewpoint. Twice I had to excuse myself to go to the WC in order to switch tapes (and to take care of ordinary business). Although I gave the appearance of having a weak bladder, it was well worth the embarrassment and slight inconvenience to be able to simply sit and listen instead of wearing my fingers to the bone by taking hurried and mostly incomprehensible notes.

After dessert I dawdled over my tea – four cups of it (but the cups were small). This overindulgence made trips to the WC more of a physical necessity than a mechanical one.

Finally, the professor completed the overview of his work. "Tomorrow I can give you some practical demonstrations in the greenhouse and laboratory."

Arlene jumped in with, "I look forward to it."

I smirked. "Well, thank you for a most enjoyable evening, Professor Wayward."

"On the contrary, I should thank you. I cannot remember when I have been allowed to prattle on like this. For a long time I have had these wonderful thoughts in my head, with no way to communicate them to anyone who would fully understand. For a layman, you have a profound grasp of scientific principals."

"That's my job." As I stood up, I said, "I thoroughly enjoyed the evening." I turned to Arlene. "Miss Hawkins." I overemphasized the "Miss." I was still pretty miffed at her.

"Good night, Tim." If that didn't sound nonchalant, I don't know what does. But I realized that she was only putting on an act for the professor. She acted as if we had never had any differences. She was a perfect lady about it, and in comparison it made me appear childish. I blushed.

Charles broke the sustained silence. "Well, it's off to the sack for me." He left.

For several dreadful seconds there was another dead silence. Then the professor broke it by saying, "Breakfast is at eight-thirty, Mr. Baker."

"Thanks, Professor. I'll be there. Er, here. And now

if you will excuse me."

"I will gladly accompany you to your tent, Mr. Baker. I must get something from the laboratory. Excuse me a moment while I give breakfast instructions to Marie."

I nodded and went out the door. I was starting along the path to the tents when Arlene called out, "Oh, Tim, could you wait a minute please?"

I stopped and turned around. I tried to act unconcerned. "Yes, what is it?"

She slipped her hand into my pocket and pulled out a roll of tape. "I just thought I'd let you know that your little trick didn't fool me. Good night, Tim. I trust you'll sleep well in your warm and comfortable sleeping bag." She returned to the house and smiled over her shoulder as she closed the door.

That made me feel as small as a toad. I think she actually took pleasure out of belittling people. Well, I'd show her.

The professor appeared a moment later. "Come, Tim. Don't let her bother you. She is acting under a common female impulse. She is completely harmless."

"Except toward morale," I replied.

"I wanted to talk about that with you, Mr. Baker. I do not actually have to get anything from the laboratory. You see, I can detect the hostility between you two. Do you have any idea why she reacts that way toward you? That is, if you care to talk about it."

"To be quite frank, Professor, I *would* like to talk about it. She told me what her problem was. She's under the impression that I'm trying to take her job away from her. Because I'm doing an article on you for a different magazine."

"I suspected as much."

"But our articles have different slants. Hers is about your involvement with the Aborigines; mine is about your scientific studies. Herbology, I believe you called it."

"Quite right, Mr. Baker." Professor paused thoughtfully. He squinted his eyes, pursed his lips, and stroked his close-cropped beard. "Well, I am going reveal a little secret which Miss Hawkins made me promise not to tell you."

At first I didn't know how to respond. "Well, don't leave me hanging, Professor. Tell me what it is."

"I will tell you only if you promise not to mention it to her."

"You're breaking *your* promise. What makes you think that I'll keep mine?"

"Because what I have to say to you might prevent you from feeling, shall we say, slighted."

Now it was my turn to squint.

"You see, when Miss Hawkins descended in that parachute she was as scared and as sick as a child. She went into convulsions, vomited, and was totally disabled for the first day of her visit. Yesterday she was still a little queasy. I daresay that she will never fly with an Aussie barnstorming pilot again."

So the Amazon lady wasn't as tough as she let on to be.

"But why are you telling me this?"

"Because she is young and impetuous, and does not know herself. I perceive that she has been making you feel absolutely miserable, and after all, I *am* a man; and I do stick up for my sex."

"Well, thanks a lot for the enlightenment. I promise not to use it against her the next time she jumps on me."

"Fine. I knew that I could trust you. Good night, Mr. Baker."

"Good night, Professor."

Professor Wayward ambled back to the house. Charles' tent was buttoned up tight, and there was no light showing through the canvas material. I had intended to begin transcribing my tapes, but found that I was way too tired to do so. Suddenly I could hardly keep my eyes open.

I didn't even look for the flashlight (or torch, as Charles called the instrument). I threw back one flap and collapsed on the sleeping bag. I was asleep in seconds.

Interlude 2

The Mold grew. The solution with which it had been soaked held the exact formula that was needed to accelerate its rate of growth to a rapid and steady geometric increase. It fed upon the piece of bread for a long time, but when that was gone there was no sustenance remaining for additional growth. Desperately in search of food, it sent out an exploratory strand between the threads of the jar and the lid. This single strand, this solitary tendril, explored tenuously, feeling, searching for food for more nourishment which it needed to expand beyond its present size.

The lone tendril alighted on a deserted bread crumb, swiftly engulfed it, and devoured it. The Mold then knew that there was sustenance outside the jar. It must get to that sustenance. But it would take too long for it to send out strands one at a time. It could subsist on itself for a while, consuming no longer needed portions of the mass that was stuck inside the jar.

In this manner the Mold could move in a linear fashion. It could grow in one direction while it metabolized unneeded portions of its opposite end. But such near perpetual motion could not possibly go on forever without some source of external replenishment.

The Mold set its many strands opposite each other and, summoning all its strength, pitted its mass against the fragility of the glass jar. It strained without success for many minutes, draining its energy uselessly. Yet during all that time the jar showed signs of weakening. Slowly it bulged outward. Then a hair-thin crack developed. The glass bulged more, but not enough to shatter.

The Mold pushed harder and harder. It would have heard a splintering sound if it possessed some sense of perceiving atomic vibration. It had no ears, or even a rudimentary tympanic vibratory organ. So it did not hear. At last the jar spread farther than the elasticity of the glass allowed. Cracks appeared up and down the side of the jar.

The jar split open and the Mold tumbled out.

The Mold had won its freedom. It reached out and greedily enveloped strewn breadcrumbs. In its wanderings it chanced to touch a half slice of bread. Slender tendrils punctured the bread like tiny darts. Once a strand was surrounded by food, that strand was able to consume the bread through the entire surface area. The Mold's rate of consumption increased manifold.

By alternately expanding and contracting its cells, it pulled the main portion of its body across the metal shelf, and implanted itself there. From there it could make a good beachhead for further progress.

Chapter 7

Except for a sore back from sleeping on hard sand, I felt completely refreshed when I awoke at first light. I rolled over onto my knees, stood partially erect, placed my hands on my hips, and gyrated my pelvis in order to work out the kinks in my lower spine. When the pain eased up, I stumbled through the tent flaps into the cool outdoors – or outflaps, as the case may be.

The sky glimmered black to the west, purple overhead, and light blue to the east. Sunrise was yet a half hour away. A few of the brighter stars shone dimly overhead. The air was crisp and clear, auguring another cloudless day. I stretched and inhaled deeply for a couple of minutes. Goosebumps rose on my arms and legs.

I stepped back inside the tent, donned shorts and sneakers, then went out for a predawn jog. That was how I normally kept my sanity and my cardiovascular system in check. I jogged eastward away from the village across loose sand and hard-pack, and around patches of shrubbery. It sure beat running on concrete, inhaling carbon monoxide fumes, and dodging cars whose bellicose drivers believed that pedestrians had no business in being on the streets.

I started at a slow pace and gradually increased my speed. I felt more out of breath than usual, but once my second wind kicked in, I was able to proceed at nearly my normal rate. The soft sand provided a welcome cushion. When my sneakers sank into softer patches, I felt as if I were wading through syrup (or treacle, as the Aussies would say). I ran much faster on the hard pack.

I ran until I saw the top rim of the sun peaking over the horizon. Then I turned and followed my tracks back to the village. The MacDonnell Range was actually a pair of ranges. The village stood closer to the southern range. I felt as if I were running along a groove in a Brobdingnagian washboard. I opined that there was sufficient subject matter for another article: the hidden beauty of seldom seen Australia.

My T-shirt was soaking wet when I returned to the village. As I walked about and stretched, Aborigines began to appear and commence whatever it was that they did for the day. They took little notice of me.

I saw Charles standing in front of my tent. I waved and ambled toward him when I was almost bowled over by two running figures in the shape of teenage boys. I shook my head in surprise, and gazed after them. They ran directly to the schoolhouse and quickly disappeared through one of the doorways. I found it odd that they were so eager to go to school – especially as class did not begin until nine o'clock.

"Good morning, Mr. Biker."

"Good morning, Charles."

"I see you like to keep in trim."

"I'm training for a marathon." I pulled my sweaty T-shirt over my head. "I thought the fresh air would improve my gait, but I had to struggle all the way."

"It's the elevition, Mr. Biker. We're two thousand above sea level."

"Ah, I see. The land is so flat that the elevation isn't noticeable. It's deceiving."

"You'll get used to it in a few diys."

"Uh, pardon my ignorance, but I forgot to ask the professor what bathing facilities are available – if any."

Charles laughed. "Outdoor shower right behind the generator shed. It'll be cold now. The water is heated by the sun. It's better to shower in the afternoon."

"I'm so hot that it won't make any difference."

I grabbed clean clothes, a bar of soap, and a towel from my suitcase in the capacious canvas tent. I was wrong about the difference. The water in the tank atop the roof was ice cold. When I pulled the cord to let the water fall through the nozzle, I inhaled so sharply that I nearly sucked in my teeth. After soaping up I rinsed off rapidly. By then I was so chilled that even my goosebumps had goosebumps. I dressed and stepped out of the wooden stall.

"Invigorating, eh, Mr. Biker." Charles was smart

enough to wash only his face and hands.

"I'll remember that about afternoon showers."

"I'm sure you will, Mr. Biker."

Charles accompanied me to the house for breakfast. He entered without knocking. I was right behind him. Marie smiled at us, and indicated chairs at the rough-hewn table.

A moment later, the professor emerged from his bedroom. Still buttoning his shirt, he said, "Well, well. Good morning, Mr. Baker. Charles. I see that you are just in time for breakfast."

Charles grinned. "When it concerns food I'm always on time."

Marie brought in plates and utensils, and set the table. I was smacking my lips and picking up a fork when she placed before me an enormous egg. It was about four inches long and three and a half inches wide. It sat in a huge earthenware cup, which took the part of an improvised egg stand.

"Do not be afraid of it, Mr. Baker," said the professor, seeing my amazement.

"Is this an ostrich egg?"

"Ostriches live in Africa. This is an emu egg. They are quite palatable, but don't feel ashamed if you cannot eat it all."

"Sure thing," I said.

"Feel free to come inside any time of the day. Just because I can not give you a room does not imply that you are not my guest. Consider yourself a . . . a . . . "

"An outpatient," came a sarcastic voice.

I looked up and saw Arlene emerging from the bedroom. No, not the professor's.

"Well, if it isn't the outer side of inner sanctum," I replied.

She ignored my statement and sat down resolutely. For the smallest instant a startled look appeared on her face, but she quickly hid it and without saying a word broke open her hard-boiled emu egg. She obviously had not seen one before.

"Good morning, Miss Hawkins. How are you?" Naturally that was the professor. I couldn't be so kind.

"Good morning," she said cheerfully.

"I trust you had a warm and comfortable sleep, Miss Hawkins." I meant to be sarcastic. Then I remembered that the professor knew of the conflict between us because he gave me a slight smile on the side.

The meal was eaten mostly in silence. I was afraid to say anything for fear of being rebuked by Arlene in her present mood. And the professor was correct about the egg. I ate only that and a piece of bread that I washed down with a cup of coffee, and I thought that I would never eat another thing in my life; but I finished it all. It was the first time I had ever gotten filled on a single egg. Arlene, of course, declined from even attempting to finish hers, and had to excuse herself.

When the meal was over, the professor said, "If you will excuse me, I have to be running along. School starts at nine o'clock sharp, and I have to ring the bell. You can browse around if you want, and if you can be at the schoolhouse by four o'clock, I will give you a guided tour of the laboratory, such as it is. Oh, and by the way, I told Marie to have lunch set for you, because I have some work to finish in the laboratory. She is very prompt, so be here at one. Good-bye."

He walked out without giving anyone time to say a word.

Charles followed him out the door.

This left Arlene and me alone. I had just opened my mouth to speak to her when Marie came in to collect the dishes. I leaned on my elbow until she returned to the kitchen, then spoke up in a commanding voice:

"Arlene – Arlene." I found myself stuttering again. Then I pulled myself together and said bluntly, "Listen, I've said this before but I'll say it again: we're going to be working together for a while, and I don't like expending half my energy in fighting you. The other day you said that you wanted to be friends."

"Don't listen to what I say. Understand what I mean."

"What's that supposed to mean?"

"It means what it means."

"And what's *that* supposed to mean?"

"It means that we can act civil toward each other."

"I've never acted otherwise."

"Well, not face to face. But you're still trying to take advantage of me."

"How?"

"By trying to scoop me."

"If anything, I would rather scoop you *up*."

"And what's that supposed to mean?"

"It means what it means."

"And what's *that* – " She stopped in the middle of parroting my phraseology.

I remembered the professor's wisdom about not acting reprehensible toward her. I could hardly believe her. One minute she was the sweetest little female that ever walked this Earth; the next she was snapping my head off. She's like that, this girl. "Does the word wishy-washy mean anything to you?"

"Wishy-washy is two words."

"Whatever. If you don't want to be friends like you said the other day, that's all right with me. But at least stop acting childish and making snide remarks. Just keep out of my way. That's the least you can do."

"All right then. From now on consider me an old friend you never got to know very well, and we'll keep it that way."

"You're not old."

"You know what I mean."

"Okay. You are now my longtime friend."

"I accept, on the condition that you don't take advantage of my kindliness."

"Agreed. It's settled." Then, "For now, let's go have a look at the sights around here. Explore the village. Talk with the Aborigines. Take some pictures of the scenery. The landscape is beautiful around here."

Suddenly she spoke with enthusiasm. "Yes. Let's."

The mysteries of the schoolhouse would have to wait until four o'clock. The mysteries of Arlene would probably have to wait forever.

*   *   *

Charles joined us on our tour of the village. Arlene and I sparred lightly while asking questions of the Aborigines. Charles translated for those who did not speak English, or whose meager command of English was difficult to understand. I found the villagers interesting if not informative: they had no information that contributed to my article. I kept voluminous notes nonetheless. As did Arlene. We both took a number of photographs of village life. She had difficulty in rewinding the film in her Nikon S3M rangefinder camera because the spindle kept jamming. I gave Charles a conspiratorial tilt of my head.

By midday the heat was practically unbearable. I stripped to shorts and T-shirt, but the bounds of decency prevented me from going any farther.

After a delicious lunch, prepared by the professor's girl Friday, I lounged in my tent and transcribed my tape recordings from the night before. By the time I was done, the recorder needed a new set of batteries, which I inserted. I could hear Charles nearby, doing mechanical work on the Land Rover.

Like the professor had said, he was ready at four o'clock sharp to show off his small classroom and attached laboratory.

I made the mistake of standing in front of the schoolhouse door when the professor dismissed his students. Consequently, I was caught in a deluge of running children. I was carried a full twenty feet by the mainstream before I was able to free myself.

The professor appeared after the children had departed, welcoming us and inviting us happily inside. "Well, well. Come right in, my friends."

"Thank you, Professor Wayward," flirted Arlene.

I simply nodded in agreement.

We entered a hallway with five doors, through one of which we had just entered. There was one at the opposite end of the hall, two on one wall, and one on the other.

"Before we begin talking, I'd like to show you

around the building. It's quite small so it shouldn't take long." He guided us through the first door to the left. Beyond was a classroom of moderate size, with armchairs in neat rows, creating perfectly straight aisles. "I teach the older children here, and they keep it very clean. Now over here – " He practically ran into the hall and down to next door on the same side. He only opened it momentarily, enough for me to get a vague glimpse of the inside. " – we have the children's room. They, of course, are not as neat as the older students in the school."

In short, without euphemizing, the place was a literal mess. The chairs were in a state of total disarrangement, books were strewn around the room, paper covered the floor like a blanket, and the wastepaper basket was empty.

"Yes, I can see that," I said, sparing the professor from any outbursts of emotion which might lead him to think that he kept a sloppy school.

Again at a maddening pace he shot out of the door, and took the few steps that separated us from the room on the opposite side of the hall. Arlene gave me a quizzical glance; I shrugged my shoulders in return.

"This is the auditorium. It doesn't look very big, but it will hold the entire population of the village's one hundred and fifty-seven individuals – standing room only, of course."

I got a fleeting glimpse of a large room cluttered with chairs before he closed the massive door. There was an outside exit on the opposite side, and I think there was also a podium.

"Now," he said furtively. "Now I will show you my laboratory. I daresay that it is most complete for my kind of work. It is filled with all the latest scientific testing material and supplies, all of which was liberated from, er, that is, which was leftover from the closing of the government testing facility. I even have an asbestos suit for descending into boiling hot springs for the much sought-after species of herbs which grow only in that small portion of the biosphere in which the temperature is such that any other kind of life form would be dead in only a few seconds."

He was obviously very pleased with this, and was most eager to show it off. I thought that I would get to see at least some part of the building halfway decently. He opened the steel door, which had a thick plastic window in the center at about head height, and ushered us inside.

"This room, of course, was specially reinforced to sustain the weight of the heavy benches and equipment. I've got bottles filled with formaldehyde and various weed killers, because I sometimes conduct some daring experiments which just might get out of hand."

The laboratory was the stereotypical Frankenstein movie set: test tubes containing multicolored substances filled dozens of racks around the room; beakers bubbled with solutions whose names were strange and unpronounceable by anyone without a degree in organic chemistry; conveyer tubes carried liquids from one glass container to another, stopping off at ill-shapen contraptions which somehow converted their chemical composition in such a way as to render them into different substances; retorts lay scattered about seemingly haphazardly, but in actuality with great purpose in mind; petri dishes overflowed with abnormal growths.

The only thing missing was a Van de Graaff generator, with its coruscating sparks of static electricity. I did not expect to find all that fancy laboratory equipment way out in the outback of Australia. The lights flickered for a moment when a motor kicked on to operate an exhaust fan.

Arlene took the initiative. "Professor, tell me, what benefit has your work in rare herbs done?"

I winced. I hoped she could write better than she could speak.

"So far my contributions have been small. I have distilled an antidote for specific types of snake venom – the king brown and the tiger snake – but I cannot take credit for the discovery. The Aborigines have known about it for generations, and led me by the hand in the process of manufacture. I *can* take credit for informing the proper medical authorities about the antidote. The Aborigines also showed me how to make an antidote for the bite of the funnel-web spider.

"The primary direction of my research is to learn from the Aborigines what they know about the medicinal properties of certain herbs, then distill the essences of those herbs and isolate the chemicals that are responsible for the curative powers of the herb. What I have found most often is that too much distillation destroys the analeptic value."

I asked, "You mean that you break down the substances so far that the constituent chemicals become harmless?"

"Precisely."

"In other words, a molecule is greater than the sum of its atoms."

"I would not have thought to put it in such terms, but you are essentially correct."

"So what's the solution." As an afterthought, "No pun intended."

The professor grinned. "The solution is to find the right solution. Many interactive chemicals require a catalyst to facilitate their combination. My research has been in the isolation and quantification of active herbal extracts.

"Not all the extracts that I have distilled have been beneficial to health. A single dram of one particular substance has the capacity to kill a mouse in less than ten seconds after ingestion, by paralyzing the nervous system. It makes a very effective mouse poison, but that is not the object of my research."

The grin vanished. "On the other hand, I have made some minor discoveries without the help of the Aborigines, some useful, some not. The work is intriguing, fascinating, perhaps even tantalizing – when I think I am close to a solution of a problem. Here, let me demonstrate one of my concoctions. This serves no real purpose other than to startle and to mystify, and it is quite harmless." In aside, "It scared me awfully when I first experimented with it."

The professor pattered around the lab, and collected several bottles and test tubes that were filled with reagents. He laid them down on a workbench and began pouring them all into a large beaker. One thing that surprised me was the fact that he did not measure specific amounts of each chemical, but just judged how much of each one to put into the solution. It reminded me of my grandmother adding a pinch of this and a dash of that to her baking recipes. Titration it was not.

After he poured five different liquids into a clear glass retort, the aqueous solution turned a bright purple. He did not light the Bunsen burner that resided under the retort. Left in his hand was a test tube containing a viscous red substance. He hesitated before pouring the substance into the retort.

Professor Wayward glanced around the lab, obviously looking for something. He walked toward a cabinet, and said, "Now where is that – ?"

He placed his hand on the doorknob and twisted it slowly . . .

Interlude 3

The Mold took perhaps fifteen minutes to devour that chunk of bread. When every last crumb was consumed and metabolized, the Mold was much larger and even more hungry. Its insatiable hunger drove it onward, made it extend its tendrils to seek more food. Growing strands outspread themselves in all directions. Somehow, it concentrated mostly on stretching laterally along the wide shelf. It seemed to sense that food lay in that direction.

A blind strand ran headlong into a jar and immediately encircled it. But the Mold's built-in food finder always pointed to that place occupied by the jar. Life-giving food was in that jar, and the Mold needed it desperately. Somehow it must get inside that jar to the food that it sensed was there.

The famished Mold sent reinforcements to the jar, while a lone strand felt around the glass for a way inside. There was none: the jar was hermetically sealed by means of a rubber gasket. Because the jar was airtight, it was also Mold-tight. But no matter what the cost, the Mold needed the food that was inside that jar.

More strands followed – dozens, scores, hundreds – microscopic in size, but wadded together to make a sizeable mass of growing, living Mold. Slowly the individual strands united, wrapping around one another until a strong fiber was formed. Then the Mold constricted itself, pitting all its strength against that of the impenetrable jar. It had made a tight rope, as thick as a mere piece of string, but far stronger.

The Mold possessed the will of an organism that had never inhabited the Earth. The many-stranded fiber coiled about the jar until it encircled it three times. Then it coupled its amazing strength with its desperate need for food, and utilized the combined powers by summoning all its strength to crush the jar.

When it had made a firm knot around the jar, it

pulled back on the loose strands, and every tendril did its job to force in the sides of the glass. With unparalleled feats of ability the mold pushed and pulled simultaneously until the glass yielded ever so slightly. The glass compressed, but did not break.

Abruptly there was a sharp crack, and broken lines splintered up and down the glass jar. With what might be called optimism, the Mold continued exerting pressure relentlessly until the jar uttered a final gasp. Then the jar imploded. Glass shot inward, passed through shards that flew from the opposite direction, and spread in all directions over the shelf.

Several of the strands were severed, but that did not matter, for each piece possessed a life of its own. Some of the strands reformed with the original organism. Others maintained their individuality, perhaps to join the mother Mold later.

Hungrily the Mold crept, or stretched, or grew into the jar and met its newfound food. It was another mold, living on a piece of bread; but this mold was small and sickly. The greater Mold fought it. The battle was short and the lesser mold succumbed quickly.

The greater Mold, once it had conquered the lesser mold, did not attempt to control its cannibalistic instinct, for it was hungry for more nourishment. It consumed the lesser mold and the piece of bread on which it had made a meager start for survival. With that added strength, the Mold sent more strands even farther afield to find more food, so that the Mold could become large and powerful, and engulf everything that was alive!

Chapter 8

Professor Wayward let go of the half-twisted door-knob; it snapped back into place. At the same time, he glanced around the room, and said, "Oh, I remember now. I put it on that table." He walked toward the back of the room. He picked up a small test tube from a table that held a multitude of glass containers. The beaker held a clear liquid that looked and sloshed like water. He brought the jar to the front of the room. "This is a most interesting experiment. I think you will find it quite amusing." He chuckled softly to himself.

He poured the newly attained liquid into the beaker. Then he picked up the remaining untouched test tube.

"Watch carefully. This will react chemically with the prepared formula, and although it might look danger-ous, it is actually quite harmless."

He poured the liquid into the beaker, then stepped back cautiously. Arlene and I needed no invitation to follow suit; we stepped back in accordance with the professor. Nothing happened for several seconds. Then the different tinted chemicals commenced to swirl like a multicolored barber pole. There was a combination of a pop and a whooshing sound. An ominous dark gray cloud shot up into the air, unfolding at the top into the shape of a mushroom cap atop a slender stem. It was an exact but foreshortened replica of an atomic bomb explosion.

Arlene jumped involuntarily, although she tried to conceal the fact from me. I'm sure that my eyes widened perceptibly, and I shook and felt shudders course up and down my spine, but through a supreme effort of will I remained outwardly motionless.

The professor's reaction was wholly juvenile. He laughed raucously, and said, "I do believe I have more fun with that experiment than with any other, even if it is scientifically useless."

He kept giggling like a child, as if the experiment had been some kind of cosmic joke. After I recovered from the shock of the experiment, and the greater shock of his unexpected laughter, I began to appreciate that despite his intellectual attainments and mature attitude, he maintained the simplicity of a little boy. It made him seem more human.

After he calmed down, we got back to serious business. I asked some thought-provoking questions about his work, and he provided very sensible and scientific answers which, after that atavistic bout of humor, surprised me. I daresay that before that meeting, I did not realize the full potential of the situation concerning herbs. There are some amazing facts that only the professor knows, and many of them he will probably keep secret from the greediness of man.

"So you see," said the professor in conclusion, "I am serving a dual purpose here. Not only do I educate the indigent Aborigines, but I also pursue scientific knowledge in peace, without the hubbub of civilization constantly hovering over my shoulder to check on my progress. I have learned some things that I dare not divulge except in the strictest confidence. Yet, on the other hand – and thanks to the Aborigines – I have made some advances in medicinal biochemistry."

He seemed to place a lot of emphasis on the facts that he was serving a dual purpose, and that he preferred be left alone in order to continue his work properly. Well, if that's what he wants emphasized, I'll emphasize it for him in my article.

As we were about to leave, the professor ambled around the lab nonchalantly, checking on the progress of various experiments. From a table he picked up a key ring that held but two keys. He fitted one into the cabinet keyhole and turned it.

In explanation he said, "I do that to keep out the youngsters. There are only two locks in the village: this one and the one on the generator shed. I encourage my students to visit the laboratory, but there are dangerous chemicals in this closet that must be kept out of reach of innocent hands."

The professor put the key ring in his smock pocket, then promptly removed the smock and hung it on a hook that was screwed into the wall next to the closet door. I don't know if anyone ever said that quaint, elderly professors are the queerest men in the world, but in case no one has, I'm saying it now. I also have to say that I think the professor is one of the happiest people in the world – not because he has achieved all his aspirations, but because he is content to live without them.

Professor Wayward strolled out of the door, followed by Arlene, followed by a coquettish glance, followed by me. That girl gives me the oddest feelings sometimes – or should I say all the time?

We were walking along the hall when the outside door burst open, almost hitting the professor in the face, and two stalwart youngsters about the age of puberty dashed wildly inside. At first they seemed not to see us, and kept right on running. After they passed us, they slowed down and glanced back sheepishly, then stopped and fidgeted with their hands.

"Boys," said the professor in a mock stern voice. "Come back here."

Both boys stood still with lowered heads.

"Now what have you got to say for yourselves?"

"Gee, I'm awfully sorry, Doctor Warren, but we were in a hurry," said one.

"Yeah," said the other.

"That's no excuse to come barging in like that. From now on you'll kindly walk into the schoolhouse."

"Yes, sir." both said simultaneously.

"And by the way," the professor stroked his beard gently. "Have you folded the chairs in the auditorium as I asked?" The professor knew darn well that they hadn't, but he made it sound as if he didn't so as to draw a confession out of them.

Two blank guilty faces were all he got for an answer.

"I thought as much. Then will you kindly go and do it now?"

"Yes, sir," echoed two harmonious voices. They

retreated slowly along the hallway and stepped into the auditorium.

As we passed from the hall to the outside and down the steps, the professor explained apologetically, "Shane and Albert. Two of my brightest pupils. They will make great scientists some day."

I asked, "Where did they pick up that American lingo, Professor?"

Laughingly he replied, "Oh that. They raid my bookshelf just about every night, and they read many of the more – shall we say – loose-writing authors of fiction."

We talked "off the record," as the professor put it, on the way to the house. Those hours of talking had made me thirsty and hungry. The professor said quizzically that Marie was baking a batch of iguana entrails especially for the newcomers.

I didn't know whether to choke or lick my lips. With this somewhat dubious dish on my mind, I was far removed from the schoolhouse and its strange happenings. While I was distracted by thoughts of herbs and iguanas, something very weird and dreadful was occurring in a certain cabinet.

Interlude 4

The Mold grew strong. It found its new type of existence difficult at first, but with renewed energy it could shoot out tendrils of itself to find new food. Mysteriously it was attracted to a large jar which took up much space in the center of the shelf. Without hesitation it encircled the jar with hundreds – now thousands – of strands and used constriction like it had on the other jars. It did not sense food in the jar; rather it sensed something vaguely familiar which was even more important: something more potent and more powerful. The Mold surrounded the glass and snapped it like a cracker.

A thin, watery liquid poured out and over the Mold. It would have gotten away but the dry Mold sapped it up thirstily, and only then realized that it was the same solution which in the beginning had given it so much strength and growth potential. Eagerly it transferred the liquid osmotically to every individual cell of every strand, Now it was drenched in the stuff, and its growing ability was enhanced a hundredfold – a thousandfold – so that its rate of expansion was becoming limitless. Now it could grow forever.

Almost instantly cells began to split and divide in two. Such an abnormal growth rate had never before existed on the Earth. The Mold descended to the next lowest shelf, and simultaneously ascended to the shelf above. Growing exponentially, it methodically combed every square inch of the cabinet. It spread to the next shelves in line, and from there onward until it had reached the fathomless depths of the floor and the unspeakable heights of the ceiling. It filled every corner of allowable space in the cabinet, and then it stopped.

In all its growth it had found only microscopic specks of food. There were many jars and test tubes which it broke, but most yielded no additional nourishment. And since it consumed no more food, it was overextending itself. Its strength began to wane. Unless the Mold obtained more sustenance, it would most certainly die.

Chapter 9

Shane and Albert were finishing their work in the auditorium. After they had every chair neatly folded and stashed against a wall in an inside corner, they dragged the table that was used as a podium and pushed it against the chairs, so the chairs would not fall down like a cascade of dominoes.

They did not even sit down to rest when they were done. Instead they raced out of the auditorium and into the lab.

Albert, the taller boy, wrenched the cabinet doorknob, but the door would not open. It was locked. He shrugged his shoulders and said, "I guess Doctor Warren locked it. What do we do now?"

"We could ask him for the key," suggested Shane.

"No, he has guests so we should not bother him." Albert glanced around the lab. His eyes chanced to alight upon the professor's lab coat, which hung within reach of his skinny black arm. A gleam entered his eyes as he jumped for the smock and felt the pockets. Almost instantly upon touching the coat he heard the rattle of keys. He searched one pocket but it was empty. In another he found several rocks. In the third he found the keys.

He knew which key was needed. He separated it from the other. He inserted the key into the keyhole and twisted it. Knowing that the door was about to be opened, the other boy backed up a couple of paces to allow for the swing. This put him several feet away from the door, and out of danger's threat.

There was a snap and the door was unlocked. Now Albert twisted the knob and the door burst open. A writhing mass of entangled Mold fell out on top of the boy. Shane involuntarily jumped back and crashed against a bench. Beakers and test tubes clattered; some fell off the bench and broke on the floor.

There was a stifled scream from the tortured boy as the Mold instantly gorged itself on its newfound food supply. It thrust hundreds of strands down the boy's throat, up his nose, and into his ears and eyes. Albert's solitary scream was stifled. Within seconds he was literally being eaten alive, as the Mold dissolved small portions of flesh bit by bit, and converted the flesh to new energy.

Many feet away from the scene of horror and disgust, Shane stood traumatized. There was nothing he could do but watch and feel his throbbing heart try to beat a hole through his chest. He kicked off his sandals which had trod the Mold on the floor. He breathed hard but did not scream in terror; his lungs and larynx were paralyzed. Instead he slid around the side of the bench and backed up farther and farther away from the writhing monstrosity.

Suddenly he stopped, and looked at the open lab door. He was safe for the moment where he stood. Yet it seemed ironic that to reach ultimate safety he would have to go out the door which stood ever so close to the creeping Mold. In order to get out of this predicament he would first have to approach the object of his terror.

He did not think these thoughts consciously. He did not weigh his options for survival. He was brilliant, but nonetheless he was a child – with a child's inherent fears. What inspiration finally drove him to run almost into the Mold and through the doorway to safety was a mystery. Perhaps he was driven by instinct, or by the horror of the scene. Whatever the cause, he ran. He ran to save his life. Blind fear must have driven him like a maniac.

He raced out of the schoolhouse with stark terror close behind. By some quirk of fate he had touched the door on his way out and pulled it in such a manner that it slowly closed and latched. The frightened boy ran through the dirt yard and along the path to the professor's house, screaming loudly enough to make long-distance telephone calls obsolete.

Chapter 10

Professor Wayward was kidding about the baked iguana entrails. Gleefully he informed us that iguanas were not even indigenous to Australia. They lived primarily in the Americas; some were also found on islands in the Pacific Ocean. Marie prepared an Australian staple: leg of mutton.

"That's an immigrant stiple, not Aborigine," Charles noted with a grin. You would never know it by the way he shoveled down the meat. "We prefer kangaroo and wallaby."

Marie grinned silently as she set dishes on the table. She never spoke. The professor explained that, although she understood simple commands in English, she spoke nothing but an obscure local dialect. I saw Arlene take a mental note.

The four of us were chatting in low tones and masticating quietly when a loud squeal or scream erupted from outside of the house. It sounded like the strident wail of a banshee.

I looked dumbly at the professor for a moment. Charles wasted no time in shoving back his chair and rising to the occasion. I was only a split second behind him. Before Charles could reach the front door, it was flung open so hard that the nails that secured the top hinge were pulled partway out of the jamb.

In ran the most terrified little boy I have ever seen. His face was so contorted with fear that I didn't even recognize him as one of the pair that I had recently seen in the schoolhouse. His eyeballs appeared to be about twice normal size; the whites were like full moons.

The boy raced past Charles and me, grabbed the professor's hand, and shouted, "It's got him! It's got him! Come, Doctor Warren. Come quick."

He dragged the professor bodily out of his chair.

"Shane, what has gotten into – ?"

"*Come quick*, Doctor Warren. Come quick!"

"We'd best go," said Charles, in a voice that was too calm for the occasion, but that spoke louder than a shout and was far more forceful.

Professor Wayward nodded like a dashboard ornament with a plastic head on a spring. "Yes, of course."

The frenzied boy continued his verbal outburst as we all proceeded out of the house and across the courtyard toward the school.

The screams had attracted the village Aborigines, all of whom came running. We were balked somewhat by the professor's shambling gait. I glanced behind me. Arlene's face mirrored that of Shane's, displaying sympathetic fear even though she had no idea what was happening. Whatever it was, she must have expected it to be bad. Her hands waved wildly in the air as she ran.

We reached the schoolhouse lickety-split. Shane halted in his tracks as if he had run into a brick wall. He extended his right arm and pointed with a trembling index finger.

"In there! In there! Albert . . . " Shane would go no farther.

Charles beat me to the door. The hallway was empty. Charles slammed open one of the classroom doors. I ducked into the auditorium. Charles slammed open the other classroom door. Then we both charged for the lab. Charles yanked open the door and stepped partway inside.

"What in the nime of – " His voice was a mere squeak.

I peered over his shoulder. Whatever I expected to see, it was worse. I gasped at a mass of pulsating fluff that occupied one corner of the room.

When Charles turned to look at me, I saw the frightening contortion of awe and alarm on his face. "What *is* that?" he shrieked in a whisper.

"I – I don't – "

I heard the professor's footfalls in the hallway behind me, and Arlene's behind his. The barefoot villagers stayed outside with Shane.

I stared unbelievingly at the white – whatever it

was. It looked like a vast bulk of cotton lint whose fibers waved in constant motion as if wafted by the wind. But there was no wind in the lab. The formless structure grew out of the tall cabinet and extended more than eight feet along the floor, writhing incessantly. Thin translucent tendrils shot up into the air like the tassels of a party blower.

Professor Wayward gently pushed me aside. "My word – " He gripped the jamb for support.

Even as I watched, the fat shapeless lump stretched forward and sideways, growing longer and wider but flatter. It reminded me of a giant hairy leech.

Arlene stuck her head past the jamb and peered under the professor's arm. She stifled a scream with a sudden intake of breath. Her hand went to her mouth.

I placed a firm hand on Arlene's shoulder – as much for my comfort as for hers.

Shane crawled through the doorway on hands and knees, between the professor's right leg and the jamb. "Albert!" he whimpered.

I perceived that the little boy was jabbering something indiscernible. I could not understand what he was saying. It was not until Professor Wayward spoke in the same indiscernible language that I realized that they were speaking an Aborigine dialect. The boy was too scared to remember his English.

A thick wad of the white fluffy mass protruded from the middle of the main body. It groped along the floor like the pseudopod of a giant amoeba. When the end of the projection touched the wooden leg of a workbench, strands extended around the leg. I heard a nearly inaudible hissing sound, like the whistling of a teakettle from a hundred feet away.

Nobody moved.

I watched in utter fascination at this new phenomenon, wondering what it meant – or portended. The pseudopodial extension compressed around the workbench leg, and worked its way upward. A gap appeared in the wood. I could not comprehend what was happening – what had happened. It seemed as if a section of the wooden leg had disappeared. The awful truth had not yet dawned upon me – or anyone.

Abruptly the professor gasped in English. "You mean that he is under that – thing?"

The boy nodded weakly. Broken foreign words punctuated by sobs of terror and grief erupted from the little mouth.

The professor looked at Charles and me. "Albert unlocked the door and that – thing – fell out on top of him."

Charles, of course, had understood the boy's language.

"You mean he's underneath that white mass?" I exclaimed.

Charles said, "We've got to get him out of there." He lunged into the room.

"For God's sake, do not touch it," the professor shouted.

Charles looked back, startled. Two seconds later he grabbed a push broom from the wall rack. Instinctively I followed suit with a sweep broom. If Albert lay under that writhing mass he might already have suffocated to death. Charles and I attacked the heap of twisting filaments with the rounded knob at the end of the brooms that lay opposite to the brush or banded straw.

I didn't feel anything solid until, judging by the length of handle still protruding from the mass, I struck the floor. Together we levered gobs of filaments as if they were heavy balls of cotton candy. We spread the mass as if it were hay in a loft. Whenever we cleared an open space, I saw that the planks of the floor were pitted, as if the wood had been partially dissolved by acid.

The four-legged workbench tilted as the surrounded leg disintegrated several inches above the floor. The fluffy pseudopod moved upward.

Charles suddenly stopped poking in the mass in front of the cabinet. He held up his broom handle. I saw that the handle was shorter than it had been; the end of it was missing. Clinging strands engulfed the end, and worked their way along the length of the handle.

The MOLD

Charles let go of the broom before the slender filaments reached his hands.

I swept my broom handle sideways through the main body of fluff, exposing and dislodging a naked skull from a liquefying ribcage. Fine filaments like thistledown engulfed the foreshortened end of the handle. I tossed it into the main mass of fluff as if it were on fire.

From the doorway the professor saw the skeletal remains. "My God, what is that thing?"

Arlene looked as if she were going to vomit. Shane backed into the corridor. I wanted to puke and run at the same time – anything to get away from the horror that carpeted the floor in front of me.

Defenseless against the unknown organism, I stepped backward as tendrils crept along the floor toward my feet.

Charles did the same.

We retreated to the doorway in the face of the advancing white translucent mass. I couldn't take my eyes off it.

Way off in the distance, I heard the professor speaking slowly and smoothly in English. "Tell me what happened, Shane. What happened to Albert?"

The boy shook his head as he sobbed and stammered hysterically. "Ye-ye-yesterday we – came in here to – grow some molds, like you – told us in your – biology lesson. We moistened – some bread and – put it in a jar. Then we – made another – and mixed up some – herb juices to see – what would happen. The darkest place – was your cabinet, so we – put them in there. Today we – came in – and the door was locked. – We found the key – in your pocket – and when Albert – opened – the door – it fell out – and got on him. – He screamed . . . "

Shane's voice degenerated to a series of plaintive whimpers, then to outright caterwauling. But the little fellow had a right to cry, for he had just seen his companion disintegrated by a monstrous organism.

The professor shook him, and asked, "What did you put in the jar with the mold? What kind of herbs?"

The boy shook his head. He rattled off some Aboriginal words that might have been common names for local species. He hunched his shoulders after every word.

"Metal? What kind of metal? Where did you find it?"

The boy jabbered in Aborigine, gesticulating wildly.

Shane spouted more dialect.

The only thing that I could understand was that the boys had made some kind of discovery that I hoped could be *dis*discovered (to coin a word).

As soon as the professor removed his hand from the Shane's shoulder, the boy ran away: along the corridor and out the door.

When I found my voice it was shaky. "What did he say? What happened?"

"I do not know exactly. Not from what he told me. But appearances indicate – " The professor pointed at the pulsating organism in the lab. " – that they placed a starter mold in some concoction that they created from a combination of herb juices and other plant extracts. They found a kind of slime on a chunk of metal that they discovered in the outback. If the Mold grew this much since yesterday, its metabolic rate must be remarkable, perhaps exponential. I suspect that somehow the Mold's genetic makeup has been altered, affecting the growth hormones – "

"Spare me the science lecture, Professor. What can we do about it?"

"Right on!" Charles averred.

"Well, I – " The professor stared at the mass of Mold.

The squirming tendrils scintillated in sunbeams that streamed into the room through the high windowpanes. The fluffy mass looked as if it were breathing. As more of the workbench leg dissolved, the bench-top tilted more out of level. Several beakers and test tubes slid close to the edge; two of them toppled over, and one rolled over the edge and shattered upon the floor.

" – I do not know. Certain vines can grow at an alarming rate, scaling a tree or a trellis practically overnight. But of course any plant or animal cannot increase its body mass without a source of food that it can convert to energy – "

"Professor!" I shouted. I jabbed my hand in the Mold's direction. "It looks like this Mold can eat anything that's organic – even dead tissue like floor boards and bench legs."

"Phenomenal," he uttered. For a moment he was distracted by the fantastic biological implications. Then he snapped back to reality. "That must mean that the boy – " He wasn't able to finish the sentence.

I nodded. "Not even a skeleton will be left."

"The Mold consumed – "

"Right on. So how do with kill it, Doc Warren? Without getting consumed in the process."

Glassware continued to clatter to the floor as the workbench canted ever more. Half a dozen petri dishes dumped their contents into the mass, providing food for the Mold.

"How do you normally kill mold?" Arlene pushed herself past the professor into the lab, but came no farther than the doorway. "Disinfectants. Detergents. Bleach. Ammonia. Anything. You poison it." She was keeping her head about her despite the stress and horror of the situation.

While we were palavering, the Mold had grown until it covered fully one quarter of the floor. It was consuming all four legs of one stool simultaneously, with the result that the stool appeared to be shrinking in height. Strange chemicals poured over the Mold and dripped to the floor. Tendrils of the Mold shot into the air like rockets, and made a beachhead on a tabletop. The sound of crashing glassware crescendoed.

"Now you're talking," I shouted.

"Right on! The laboratory is an entire roomful of chemicals. Some of them must be acidic or basic enough to kill a simple thing like mold."

"I daresay that this is no simple Mold." The professor remained calm despite the extremity and tragedy of ongoing events. "Nonetheless, it cannot be invulnerable – " He slouched against the jamb resolutely, despairingly, and stared at the floor.

"What can we do?" It was a useless question and I knew it, but I had to say something. I also suspected that although the professor acted like he was beaten, he was thinking his head off. He was as desperate as I was, only he didn't show his anxiety by outward emotional display.

I glanced at Arlene and saw her eyes widen in horror. She, too, was looking at the floor where the professor's gaze was affixed. She pointed toward the other side of the lab. I followed her shaking finger and saw that strands of Mold were beginning to slither under the window frame.

I touched the professor's arm. "Professor, we need to do something, and fast."

He followed my gaze and stared at the ever-creeping Mold. Slowly he stepped back, as if in disbelief. "Chemicals. Yes. Miss Hawkins is right. Chemicals will do it."

I noticed a fire extinguisher hanging on the wall by the door. "Get out of the way!" I yanked the red cylinder off its holder, pulled out the retaining pin, directed the nozzle at the Mold, and squeezed the handle.

Carbon dioxide burst from the nozzle in a thick white cloud. Charles grabbed another cylinder and followed suit. Together we sprayed a thick carpet of powder on the Mold. I had the impression that the tendrils that extended farthest from the main mass cringed or curled, withdrawing somewhat from the onslaught of cold and chemical reagent.

Then I realized my mistake. Animals need oxygen to metabolize food, and they exhale carbon dioxide as a waste product. But plants utilize carbon dioxide the way animals utilize oxygen. Perhaps we were furnishing the very chemical that it needed to increase its metabolic rate. I stopped spraying.

I grabbed Charles by the arm, and shouted, "Wait a minute! Carbon dioxide isn't harmful to plants. They thrive on it."

Charles grimaced. "It's still a chemical. A cold chemical. It might work." He stopped spraying anyway.

"Mold is a fungus, not a plant," said the Professor.

As if that meant anything to me.

The four of us watched to see what effect the carbon dioxide would have on the Mold. Would the carbon dioxide suppress its growth, or help to feed it? The extremity tendrils flayed about like whip ends. The workbench with the half-eaten leg tipped over with a crash. Some glassware broke upon contact with the floor, but other glassware was protected from breakage by the thickness of the Mold, which acted as a cushion.

"Keep spraying," said the professor. "Mold needs oxygen to metabolize."

We sprayed. We inundated the Mold with fire retardant, smothering it under a blanket of cold carbon dioxide.

In the middle of the lab, one jar that was filled with clear liquid tumbled over the edge of a tabletop. The jar didn't break, but it spilled its contents over an extended mass of Mold. Immediately, the portions of the Mold that were soaked with the solution turned black. The affected Mold sizzled with a sound that reminded me of water squirted onto a fire. The sizzle was louder than the hiss of disintegration and the din of collapsing furniture and breaking glassware.

"Do you see that?" Charles shouted.

"I see," said the professor.

We stopped spraying. The Mold avoided several areas in which chemical spills were inimical to its life processes, but of all the acids, alkalis, extracts, and reagents in the lab, that one clear liquid seemed to be doing the most damage. The blackened areas withered and died − or so it appeared. At the very least, the blackened areas no longer moved or grew. Instead, the main mass sidestepped the affected areas. I supposed that "sidestepped" isn't the appropriate word; the Mold receded from the inundated area while at the same time remote strands withdrew and reinforced the main mass. I had the fleeting impression that the Mold was healing itself: pulling itself together in order to protect its core.

"We must know what was in that jar," said the professor smoothly.

"How?" Charles and I asked simultaneously.

"Everything is labeled."

"I can't read it from here," I said.

The Mold stopped growing in the chemically affected area, but the black scar from that clear liquid solution was spreading throughout the organism like a plague. The infected parts were withering and dying.

The professor stroked his beard and assessed this new situation. "It must be an osmotic process which the Mold is unable to curtail."

"That's grite, Doc," Charles said sardonically. "Just wonderful."

The wooden flooring was showing where the clear liquid had slopped. The expansion of the dead zone appeared to be decelerating. The "hole" in the middle of the Mold was still growing, but more slowly than it had been at first. The tendrils around the perimeter of the hole were browning around the edges, but were no longer turning black. It reminded me of a burning pile of autumn leaves in which the fire was slowing dying.

"I don't know what we're dealing with, but I siy that we burn down the building."

"We will if we have to, Charles, as a last resort."

I assessed the situation and counted my chances. "I think I can jump over the living Mold and land in the dead zone and grab that bottle."

"No!" Arlene shouted vehemently. "It's too dangerous."

"Do you remember what was on that workbench?" Charles asked.

The professor shook his head in resignation.

I didn't wait to argue. I stepped back to get a running start. I just hoped that the chemical solution wasn't so rare that the professor didn't have any more in stock. Before I could take a single step, the professor gripped my arm with the strength of fierce desperation.

"Miss Hawkins is correct. You and your clothing are organic. If any Mold gets on your shoes or slacks or other garments, it will take hold and start to consume the material."

I was poised for the spring but I immediately relaxed my muscles. The professor was right. From what I had seen in the lab, the Mold was incredibly infectious. I breathed deeply as I looked from one person to the other, and at the shriveling Mold.

I held up my hands, palms outward. "Okay."

The dead zone had stopped expanding. The brown withered ends that surrounded the hole were slowly turning white, and the outer periphery of the mass was beginning to wriggle.

Four minds thought furiously for seconds that passed like hours.

Finally, the professor spoke in a voice that concealed any sign of trepidation. "Mr. Baker, do you remember me telling you about that asbestos suit? The one that I use to obtain herbs from hot springs?"

I didn't see anything burning, or the need of a fire-retardant suit unless he was planning to torch the building and incinerate the Mold. "Yes. So?"

"Asbestos is an inorganic material. It is neither animal nor vegetable, but a mineral: magnesium silicate. Its fibers are resistant to most chemicals."

I didn't quite understand where he was going with his line of reasoning. "Yes, yes, get to the point."

"Well, if someone could get into that suit and go into the lab . . . "

I hefted the fire extinguisher and bellowed in a burst, "And when he comes out we can squirt him with this or a chemical disinfectant and cleanse the suit of infecting Mold so it won't spread. That's it!"

"Charles . . . "

"I know where it is." Charles dashed out of the lab as if all the hounds of Hell were on his Aboriginal tail.

"It's in the generator shed," Doc explained.

I watched the Mold helplessly during Charles's absence. It had overcome its temporary setback and was growing again: not just outward from the outer edges, but into the scorched hole in the interior of the mass, where the chemically burned hole was filling in.

I looked into Arlene's glimmering green eyes. "That was a good idea about the disinfectant."

She shrugged. "You would have thought of it if you had been a housewife."

Charles burst into the doorway at the far end of the corridor. Draped over one arm was something that looked like a high-altitude pressure suit complete with rough-soled boots; it reminded me of a pair of child's pajamas with attached booties, and was nearly as small. In the opposite hand he held a pair of gloves and a long-necked hood that sported a clear glass faceplate.

"I don't think it will fit me . . . " Charles began. He was taller and huskier that I was.

"Not to worry," said the professor. "I can – "

I was about the professor's size. I yanked the asbestos suit from Charles's grip. "It will fit me just fine."

Before anyone could protest, I sat on the floor and kicked off my shoes. I forced my feet into the legs of the suit, then stood up and wiggled as I pulled the suit up and over my waist.

"Mr. Baker, it is my responsibility to – "

"Save it for the encore. It's no more your fault than it is the Queen of Sheba's, so stop blaming yourself. I'm not only faster and more nimble than you, I'm more expendable. If this trick doesn't work, you can always think up another way to kill the monster."

"Tim, please be careful."

"You act as if you almost care."

She touched my neck with fingers that, despite the heat, were as cold as ice. "But I do, really."

"I'll remind you of that if I survive."

"I hope you do."

" 'Frailty, thy name is woman.' " I quoted more sarcastically than I intended it to sound.

Wordlessly, Charles helped me with one arm while Arlene helped me with the other. I pulled the zipper up from the waist to my chin. Then they helped me to don the gloves and seal them on the wrists with asbestos straps.

Charles spread the neck dam of the hood and pulled it over my head. He adjusted the angle until the faceplate was positioned in front of my eyes. Then he tightened the neck of the hood around my throat with asbestos straps. Hopefully the neck seal would prevent any stray strands of Mold from entering the suit – and my person.

The resilient Mold was now recouping its strength after the partial dismemberment caused by the chemical bath. It was engulfing tables and benches with evident abandon. Glassware clattered as it fell on the carpet of white Mold. Most of the chemicals had little or no effect on the Mold, other than to dampen it. Vapors rose toward the ceiling in those places where certain chemicals reacted with the Mold. But nothing destroyed the Mold better than the clear chemical in that unbroken flask.

I peered through the fog that was slowly filling the laboratory. "I'm ready."

"Good luck, man," said Charles.

"Be careful, Tim." Arlene's tone was convincing.

Professor Wayward was silent.

Charles sprayed my suit with carbon dioxide as I advanced through the doorway. The Mold had sprawled so far from the cabinet that it swirled around my boots after I had taken only a couple of steps. I was surrounded by living, clutching Mold!

The wooden floor creaked and groaned and felt mushy. The Mold was consuming it. If I didn't hurry, I might break through the weakened planks and fall into the crawlspace. I might not be able to extricate myself. Worse, the asbestos suit might get torn in the process, and leave me open to invasion by the ever-seeking strands.

Hundreds, perhaps thousands of tiny tendrils shot into the air around my feet. They wrapped around my ankles and calves like cords. I could actually feel the Mold's strength as it constricted on my legs. I almost tripped and fell because of the unexpected attack. Then I learned that I could easily snap the slender strands by jerking away from them suddenly.

I tramped through white fluff that was now more than two feet thick in the middle. The Mold had grown so fast in the damaged area that the jar was nearly buried. The jar lay twenty feet from the doorway, but it took me the better part of a minute to reach it. I had difficulty in lifting my feet from the Mold-covered floor: not because of the weight of the Mold, but because the strands were tacky. I felt as if I were wading through syrup, or treacle.

With a thickly gloved hand I dashed away the curling strands that smothered the glass. The liquid that was inimical to the Mold had leaked through the floorboards. In places I could see into the crawlspace between the joists. The Mold was rapidly consuming the floor: boards, joists, everything but the iron nails.

I was careful not to lose my balance as I bent over to pick up the jar. Some of the death-dealing fluid was still inside. I poured a bit of the remaining liquid on the Mold that was twining around my feet, and snickered with glee as the tendrils dissolved with the speed of sulfuric acid eating through Styrofoam.

Then I turned and made my way back to the doorway through the viscous carpet of Mold. As if the Mold could sense my presence and proximity, it shot strands at me unerringly from lofty tables and benches. Tendrils snaked around my arms. The stuff was pulling at me from all sides. I felt as if I were in a living snare. Wisps grabbed me around the neck, others wrapped around my arms and legs, even around my waist. It felt as if a myriad tiny ropes were lassoing me, like Lilliputians roping Gulliver and pulling him to the ground. I resisted furiously. I yanked off the grasping tendrils like one does to a nest of hornets which have just landed on his face. Strands draped down over the faceplate. I wiped the glass clean, feeling averse terror coursing along my spinal column.

The feeling of urgency welled within me. I feared that the slender strands might poke through the seals of the neck and wrists. I wanted to run, but couldn't.

My feet felt like dumbbells when I tried to lift them clear of the fluffy mass. The Mold was gaining some indomitable strength along with its increasing proportions.

I pulled free from the writhing multitudinous entanglements. A blast of white snow encircled my face and chest, abdomen and hips, upper and lower legs. I wiped the faceplate clean, saw Charles spraying the floor around my feet. I stepped clear of the Mold but was careful not to enter the hallway or touch anyone until Charles completed the decontamination procedure.

Charles pulled me into the corridor and slammed the door shut behind me. I fumbled with the neck strap, then pulled off the hood. I was breathing heavily in the thick atmosphere, and I was sweating profusely. Charles played the fire extinguisher around the doorframe.

I held out the jar to the professor. He took it gently.

Arlene wiped sweat off my face with a red bandanna that she pulled off her slender pink neck. She gave me a firm hug that I was too debilitated to appreciate. I felt as if I had just run a marathon instead of walking twenty feet across a laboratory. Charles and Arlene helped me to strip off the asbestos suit.

After I could breathe normally, I asked, "Can you tell what it is, Professor?"

He snickered as if his laboratory wasn't being overrun by a monstrous Mold. "It appears to be a concentrated saline solution." Without further ado he put the jar to his lips, tipped it over, and poured a dollop into his mouth. He worked his tongue back and forth for several seconds. "It tastes like ordinary seawater."

All three of us chorused. "Seawater?"

The professor stared at the jar. "Part of a sample that I procured from the ocean during a recent holiday to the beach. I was collecting oceanic microorganisms for analysis and study." He looked at each of us in turn. "Some molds grow in seawater, but this particular species must find the medium inimical." He stared calmly at the ceiling as if he were speculating about a scientific experiment. "The killing agent might be sodium chloride."

"You mean salt?" I exploded.

He nodded. "I would have to conduct some tests to be certain, but it is possible that the ionization potential attacks free radicals in the Mold's metabolism. When the liquid medium is absorbed – "

"Common table salt?" Arlene wanted to know.

Again the professor nodded. "Potassium chloride would likely have the same effect, as might any crystalline compound which ionizes in water, or which – "

Charles interrupted. "If that's what kills the Mold, can't we mix regular salt with freshwater and throw it on the Mold to kill it?"

"I do not see why that would not work. Of course, other reactive solutions might – "

Charles waited to hear no more. He started running, and I was right on his heels. We burst out of the schoolhouse to where the villagers had gathered. They were jabbering plaintively in their lingo. Charles shouted with electrifying effect; the general hubbub ceased.

While he was eliciting their aid, I dashed into the house and grabbed the saltshakers off the dining room table. I did not see Marie in the kitchen. I presumed that she had joined the mob in front of the schoolhouse. Quickly I threw open the cabinet doors, searching for a bag, box, or carton of salt in bulk. I found two one-pound cartons that had not been opened, and one that was partially filled.

By the time I reached the schoolhouse with my booty, villagers were returning from their huts with buckets and pails and jugs and teacups: anything that would hold water.

Charles saw the salt containers in my arms. "We've got more coming, Mite. We use it for curing meat."

I heard a crash from the side of the building, and the tinkling sound of broken windowpanes. Arlene and the professor emerged from the schoolhouse.

"It's broken through the glass," Arlene explained.

Professor Wayward was holding a bucket with a

garden sprayer. "We can mix our saline solution in this, and spray it on the exterior walls and windows."

In the near distance I saw a wizened old man operating a hand pump at a well head. The vanguard of a bucket brigade was already arriving with cool ground water.

"How much salt do we add?" I asked.

"Brine is generally about a three percent solution. I could work out the precise concentration on paper if I knew the volume of the buckets," said the professor. "But I can do it faster by taste."

A villager slopped some water into his garden sprayer bucket, then raced back to the pump. The professor took some of my salt and dumped it into the bucket, stirred the water with his hand, scooped some out with his cupped palm, and licked the water off his skin.

"Another dash should do."

Thinking of my grandmother's improvised recipes, I added a dash of salt to the bucket of water. Charles grabbed the bucket and ran around the side of the schoolhouse to where the Mold was breaking through the windowpanes. He placed the bucket on the ground. Holding the sprayer in one hand and the pump handle in the other, he sprayed liberally around the window frame. Instantly the white fluffy Mold turned black, withered, and died.

From a female villager, Arlene took a canvas sack that was filled with salt. She threw a handful of salt into a pail of water, and stirred with her hand the way the professor had done.

She sampled the final mixture. "Tastes like seawater to me."

I took her word for it. I carried the pail into the schoolhouse and along the corridor. When I opened the door to the lab, I saw that the Mold had grown right to the sill and was crawling up the jambs. I sloshed some saltwater on the floor in front of me, then scattered the rest onto the twitching white fluff.

The Mold instantly hissed like escaping steam, and

turned black. A wave of relief came over me when I saw that the salt solution worked as directed. My relief didn't last, for I suddenly felt naked and defenseless with an empty bucket in my hand.

A moment later a long line of villagers – men, women, and children – started tramping into the schoolhouse, all bearing containers of water.

Arlene waved from the doorway. "It's already mixed."

I caught a glimpse of the professor outside, hunched over a bucket into which he was sprinkling salt crystals.

The villagers handed the buckets to me, waited while I emptied them, then ran away with the empties: a Mold brigade instead of a fire brigade. When I worked my way into the lab, I saw Charles smashing in the window with a balk of timber. He sprayed the frames until the wood was saturated, then commenced to throw in water by the bucketful.

And so it went: the villagers ran back and forth carrying an endless stream of water; the professor and Arlene added salt to create artificial brine; Charles and I tossed the solution onto the Mold. I prayed that we had enough salt to do the job.

It was the weirdest battle I could imagine. I didn't think that I could ever take another shower without being frightened out of my wits by a scum of mold on the tile, or see a slice of moldy bread or fruit without gagging in horror. We all worked ceaselessly for hours.

Long after every strand of Mold had been dissolved, we continued to pour water into the laboratory. Water was the universal solvent. We saturated the floor, the tables, the benches, and then went to work on the walls and ceiling. Nothing of any value was going to be salvaged from the room. After every surface in the lab was soaked, we soaked it again.

And again.

And yet again.

It was a long, hard, and weary fight, but in the end we won.

Interlude 5

There was a vast flood of toxic liquid, and the Mold could do nothing about it. There was no way out, no way to fight for survival. Yet the ravenous Mold was still hungry. It was dying.

Silently it did all it could do. It sent out shoots like always, hoping that the deluge would stop, for it was rapidly losing its newfound hold on life. By chance it happened to send a single, lone tendril through the floor and into the foundation. Because the foundation was constructed of cinder block, it held many holes that were big enough for the tendril to enter. The lone strand sneaked through the pores and found a vast abyss beyond. The cinder block was hollow. The Mold found a place of refuge.

Suddenly there was a splash of ionized water. The poisonous deluge dissolved the outside portion of the Mold. The single strand separated from the main body.

This lone dissociated strand needed food in order to thrive and regenerate. It sensed a few chips of wood in the bottom of the cinder block hole. The strand swooped down on them and with its last remaining strength engulfed the wood.

At the bottom of the hole the Mold might manage to persist. That infinitesimal piece of Mold, which had once been part of a big and powerful organism, now struggled to survive: to cling to that phenomenon called life.

It was a solitary desperate strand. But it was alive.

Chapter 11

That night I slept like the dead. I didn't even mind the rocks under my sleeping bag, or the chill that crept into the tent. I was afraid that my dreams might be filled with monsters – creeping, suffocating monsters – but if I had any dreams (or nightmares), I forgot them as soon as I opened my eyes in the morning.

The sun was already high in the sky. I heard villagers going about their morning chores, as if nothing untoward had occurred the day before. I reached for my jogging shorts, but my aching muscles put paid to any thought of taking a constitutional. I lay there like a lump, unable to move.

"Morning, Mite."

I groaned as I rolled over and flipped open the tent flap. A bright yellow orb burned into my eyes from the horizon.

"Quite the adventure, eh?"

I groaned again.

"Are you hungry? Breakfast is already on the tible."

I didn't feel hungry but I figured I could eat. "I'll be along in a minute."

"Right."

After Charles departed, I lay there for a full five minutes, trying to wipe the cobwebs from my head. It didn't work. So I rose, dressed, and stumbled to the shower. I couldn't stand the thought of cold water on my body, so I just splashed some on my face to wash the sleep out of my eyes.

The morning mood was somber. Marie ladled steaming porridge into carved wooden bowls. Arlene barely nibbled. The professor played with his food with evident contemplation. Charles wolfed down his meal as if it were his last supper before crucifixion. After inhaling the aroma, I found that I was hungrier than I expected. The clatter of utensils sounded loud in the silence.

The professor was the most dejected-looking man I ever saw. "I have suspended classes for the day."

I nodded perfunctorily.

"Albert's parents – indeed, the entire village – are in mourning."

I thought of a sick pun about mourning in the morning, but kept it to myself. It didn't seem funny under the circumstances. "How is Shane taking it?"

"His parents will not let me see him. I suspect that he is in shock. Nothing like this has ever happened before."

"I hope it never happens again."

I glanced at Charles, thinking that he might say "Right on." But he was grimly silent.

Arlene shuffled her feet. She took a small bite of her porridge.

I tried to break the mood. "Any idea, Professor, how that Mold came to be so – vicious?" When he didn't answer right away, I asked questions that were more general in nature, hoping to pry him out of his introspection. "What exactly is mold? I mean, where does it come from? How does it grow? How does it reproduce? I thought mold was a plant, in the plant kingdom, but yesterday you said it wasn't. So what is it?"

Professor Wayward toyed with his porridge as he pondered my rambling list of questions. I could almost see the wheels turning in his head. "Molds and mildews are primitive microscopic fungi of astounding diversity. They are separate from – "

"Fungi? You mean, the plural of fungus? Like mushrooms and toadstools."

I thought of the grade school ditty, "There's a fungus among us," but kept it to myself. Ditto with, "There are fungi in my eye."

"Quite right, Miss Hawkins, although mushrooms and toadstools are macroscopic species. Toadstools are differentiated from mushrooms by their inedibility; they tend to be poisonous in varying degrees to humans."

"So they *are* plants?"

"No. Plants synthesize carbohydrates from carbon

dioxide and water, using light from the sun as a source of energy – a process that is called photosynthesis. Fungi are quite different in that they derive their energy for growth and reproduction by secreting enzymes that decompose organic matter, which is then absorbed and metabolized by combination with oxygen. Fungi do not need the sun, and live quite well in the absence of sunlight. Taxonomically, fungi occupy a kingdom that is distinct from those of plants and animals. In the grand scheme of life, they are parasitic in that they consume plants and animals the way animals consume plants – and sometimes other animals."

"I thought everything on the planet was either animal, vegetable, or mineral."

"That is the simplistic view, Mr. Baker, which is taught in high school biology class. Today we recognize three major kingdoms – animal, vegetable, and fungus – and many biologists argue that other forms of life are distinct and deserve their own kingdoms: unicellular organisms such as bacteria, for example; and submicroscopic viruses that consist of ribonucleic acid and deoxyribonucleic acid within a shell of protein.

"Some biologists contend that viruses are not alive because they cannot exist outside of a host cell. It all depends upon the way life is defined. Life is generally defined to describe an organism that grows and reproduces. But crystals grow and reproduce, yet we do not think of them as being alive. The definition of life is strained at the microscopic level. One might say that the true definition of life is philosophical as much as it is biological.

"But to return to molds. There are quite literally thousands of species of mold. I daresay, there is no place on Earth where molds are not found in abundance. Some species live in arctic conditions, while others thrive in temperatures that exceed 500 degrees Fahrenheit. Some species live in lowland deserts, while others live on top of tall mountains. There are aquatic species – which are unaffected by the electrical charge of ionization.

"Molds are not interchangeable, however. A high-altitude species cannot live in a hot spring, any more than a bat can live underwater. Each species is adapted to a specific niche. In many cases, that niche may be practically universal. For example, common bread mold lives everywhere in the world; that is, in all dry lands of the world; not under water – although it needs moisture as part of its metabolic process."

"How does it get from one place to another?" I asked.

"Bread mold does not travel. Although some fungi reproduce by means of budding, bread mold, like other molds, reproduces by means of spores which are emitted into the atmosphere and carried by the wind. If a spore lands on a viable food source, it divides and grows into a new organism. If no food is available where it lands, a spore may remain dormant for a long period of time and under adverse environmental conditions: desiccation, for example, and extremes of heat and cold. Mold spores are not invulnerable but they can be quite indestructible. The spores reactivate when they are touched by food. They live as long as the food lasts, and emit spores into the atmosphere to create a new generation."

The professor paused to take a spoonful of porridge, and to wash it down with a sip of milk. He seemed to be warming to his subject.

"As a kingdom of life, fungi are quite versatile. I daresay that the world would be a poorer place without them. Fungi and molds are largely responsible for the decomposition of dead matter – although bacteria play their part as well. When you see mold on bread, you are witnessing nature in the process of recycling. The bread contains energy which opportunistic mold knows how to release."

More porridge, more milk.

"One might say that molds are essential, perhaps indispensable, for maintaining life on the planet. Without decomposition, dead matter would overwhelm the world. Succeeding generations of plants would be unable to gain a foothold in the earth. Animals would have no place to live. The surface would be overcome with waste product: an enormous garbage dump in which nothing could live."

Porridge, milk.

"From the human perspective, in addition to molds that are beneficial because they promote decomposition, some molds are edible, some are medicinal, and some are toxigenic. The edible species are employed largely as processing agents, such as those that are used in the production of certain cheeses: blue cheese, Roquefort, and the like. The most famous medicinal mold is *Penicillium notatum*, from which penicillin is derived. Toxigenic molds produce mycotoxins which are harmful if ingested in sufficient quantities."

After inspecting it carefully, the professor took a bite out of a buttered bun. He was on a roll.

"Bread mold is not poisonous, but it does not add desirable flavor to the product the way cheese molds do. Perhaps the most useful fungus in human terms is yeast. There are hundreds of species of yeast that are utilized to leaven bread, age wine, and brew beer, all of which is accomplished by the decomposition of carbohydrates. In Asian countries, the mold *Aspergillus oryzae* is cultured for the fermentation of a mixture of wheat and soybean to produce soybean paste and soy sauce."

"On the other hand, fungi and molds can be a nuisance. Ringworm is not truly a worm, but a parasitic fungal infection. These dermatophytes feed on fibrous proteins called keratins, which are found in the outer layers of the skin. Other fungi cause athlete's foot, skin rashes, itching, and so on. Some molds grow on bathroom tiles and exude an offensive, sometimes choking odor. This may lead to respiratory problems and sometimes to death."

I held up my hand like a student in school. "Professor, how can mold grow on tile? Tile is ceramic, made from clay. It isn't organic."

"True, Mr. Baker, but bathroom tiles that are not

disinfected regularly become coated with skin cells and oils on which mold can feed."

"Ingenious little critters," I said, shaking my head.

"We inhale airborne mold spores with every breath, Mr. Baker."

Arlene dropped her spoon with a clatter. "Not – from *that* Mold?"

The professor shrugged. "I do not know enough about the species to make any determination.

"And I hope there's none left to study," I said with a shudder.

"Right on!" Charles declared.

"I quite agree." Porridge, milk. "Some molds reproduce by budding instead of by sporogenesis. In budding, an outgrowth of the parent organism is detached, and then forms a new individual. Other molds, such as slime molds, do not consist of cells and do not form colonies. The structure of a slime mold is called a mycelium, or plasmodium, which is basically a mass of consolidated protoplasm which is not separated by cellular walls.

"Yuck." That was Arlene.

"Like a gelatinous mass of frog eggs," Charles offered.

"Or Jell-O, or junket," I added.

"Precisely. Slime molds proceed through various states. In the vegetative state they consist of unprotected protoplasm, much like gelatin in appearance. In another state they may produce motile cells – that is, cells that have the power to move. In yet another state, some slime molds may produce hyphae: threadlike extensions or branching filaments that are interconnected throughout the structural mass. Hyphae are not cilia or flagella, which function as organs of locomotion in single-celled organisms, but are more like the thick threads that hold a hairnet together."

I winced. "I'll never put jelly on moldy bread without shuddering."

"So what was – our Mold?"

"An excellent question, Miss Hawkins. Slime molds

flow over their food in order to ingest it, much like amoebas do. Other molds send out filaments, or tendrils. From the limited observations which I was able to make under admittedly trying circumstances, our Mold appeared to do both – and with amazing celerity. The outer membrane was gossamerlike, yet the Mold also extruded hairlike strands with seeming – volition – as if it sensed somehow a source of food – and of danger."

The professor paused for a moment of silent contemplation.

"If I were to hazard a guess, I would suggest that we are dealing with an incipient organism. One that has never existed before."

Chapter 12

That day, the professor suspended all classes while the massive clean-up project was underway. The temporarily dismissed students spent their normal school hours in emptying the lab of its paraphernalia. They placed the unconsumed furniture in a haphazard array for inspection and eventual reconstruction. The older children and some of the adults set about to cut and carve new legs to replace those that the Mold had consumed.

Charles, Arlene, and I helped to clean and disinfect what remained of the lab. Needless to say, the place was a shambles. As if the water damage wasn't bad enough, the salt crystallized after the water evaporated and settled on everything, either infiltrating it or corroding it.

The professor ambled about like a lost soul. His beautifully equipped laboratory, which was worth thousands of dollars and which had taken him a long time to build, was in ruins. There was barely a test tube that the mold hadn't crushed, nor a crevice that the salt hadn't infiltrated. But then, that had been our purpose, to get into every crack and cranny, and make sure that not so much as a single strand survived.

I suppose that as long as we had expunged the Mold, we could count ourselves victorious. But, as King Pyrrhus noted wryly when he defeated the Roman army but suffered staggering losses, "One more such victory and I am undone." Ours was nearly a Pyrrhic victory.

We salvaged whatever equipment was unharmed or repairable, and piled the rest in a heap for later disposal. Oddly, much of the glassware survived the catastrophe, as did copper tubing and metallic apparatus. The items that suffered the most damage were those whose components were organic. The Mold had dissolved cardboard boxes, paper pads, pencils (but not the graphite), pestle handles, even rubber tubing (which came from tropical plants and trees).

The roof was intact because the Mold had not eaten through the ceiling. Charles and I tried to place a sheet of plywood over a broken window frame. When he started to drive in the first nail with a hammer, the sash fell out of the partially dissolved casement. I barely got out of the way before the assembly tumbled to the ground.

Arlene came running when she heard the crash. "Are you all right?"

I hummed a familiar tune but kept the words in my head for fear that she would think that I was acting juvenile: "I'm a mold cowhand, from the Rio Grande." It *was* juvenile, but that's beside the point. Out loud, "I've been better."

Charles said, "The outer wall needs to be replaced, as well as the floor planks and joists. And if we don't shore up the rafters soon, the roof might collapse. Hiy, look who it is."

With an ear-piercing squeal of brakes, an aged, dusty four-wheel-drive pickup truck ground to a halt in front of the schoolhouse. Two tanned, good-looking men emerged from the cab.

"Professor," I called out, "You've got company."

The professor leaned out of the place where the window used to be, far enough to see the visitors, then climbed through the opening.

Both men were coolly dressed in khaki shorts and sleeveless shirts. The taller one stood well over six feet, and was thin to the point of emaciation. He sported a dark face because of a thick and curly but well trimmed beard. The other, clean-shaven and some five inches shorter, was broad and muscular. A whopping head of auburn hair made up for what he lacked in height.

The tall one whistled, saying, "That's quite a mess you've got here, Prof."

"Did you have a fire, or did the natives revolt?" asked the other. "Don't see no burnt wood."

The professor wiped his hands on his dirty smock, and looked back at the schoolhouse and the mess that was spread on the ground. "I am afraid that we had a little accident."

"Little ain't the word for it," guffawed the shorter man who, with a wide sweep of dark brown eyes, looked the place over carefully. Then he walked to the damaged and largely nonexistent laboratory. He peered inside and uttered some kind of mild oath under his breath. "What kind ov a accident could cause all this mischief? From the looks o' things, a small typhoon formed inside the lab." Turning to his companion he said, "Come look at this, Ed."

Ed spoke in a kidding manner, "Your atomic bomb trick didn't get out o' hand, did it, Prof?"

So the professor had shown them his mushroom cloud science experiment.

"Worse than that," answered the professor.

"Yeah, we had a little problem with Mold and decided to eradicate it."

The professor turned to me, and remembered that we hadn't been introduced. "Oh you haven't met Tim Baker, a reporter from the States. This is Edward and Albert Enders. They stop by whenever they have business in this part of the outback. They are brothers."

"Nice to meet you." I shook hands with both of them.

"And this is Arlene Hawkins, also a reporter from the States." He pointed his hand at the brothers. "Albert and Edward Enders." In aside, "I switch their names every time so as not to show favoritism."

Arlene stepped forward, and nodded. "Does everyone around here know everyone else?"

"Pretty much," said Albert. "Don't get many strangers in these parts."

I closed one eye against the sun. "You fellas don't sound Australian."

Albert guffawed.

Edward turned up the corners of his mouth. "We're from North Dakota. Moved out here 'bout twenty year ago, after the fokes passed away."

"Dakota got too crowded. We like wide open spaces."

"Sold the ranch and bought a passel o' land in South Australia. Ran cattle down there till the gov'ment relocated us."

Albert leaned forward conspiratorially. "Atomic bomb tests, you know."

"An' spaceships. The noise scared the milk outa the cows. We got a spread nor'east o' here now. Cattle are fenced in by the mountains and rivers, so we spend our spare time drivin' to the outposts."

Charles laughed. "These blokes know the Northern Territory better than I do. And immigrants, yet."

Albert guffawed.

"How come you ain't warmin' your behind on a chair in the Springs?"

"I'm on safari. Mr. Biker here is my employer. He's doing an article on Doc Warren. Miss Hawkins, too."

"Well, it looks like yer gonna have a story to report," Edward said, smiling.

Albert guffawed.

"So what *really* happened?"

"It is a long story," the professor grimaced. "I suggest that we talk about it over a spot of tea."

"Sounds good, if you can change the tea to cowboy coffee."

"Marie can do that."

Marie emerged from the kitchen as we entered the house. Her ebony face was split with a big grin that contrasted sharply with her bright white teeth.

"Tea for four and black coffee for two."

Marie nodded, and ducked back into the kitchen.

We settled down on the rough-hewn dining room chairs. Soon I heard the teakettle whistling: it did not take long for tepid water to boil. Cowboy coffee was made by throwing loose grounds in the bottom of a kettle of boiling water.

Edward was the first to speak, "Now about all that damage in the lab . . . "

"Yes, well, it is quite an unusual story, but my companions can bear me out on the facts, along with the entire village, for they played a great part in the ordeal.

You can save your disbelief until the end for verification. In any case, this is the way it happened. Two of my prize students were experimenting with molds . . . "

Interlude 6

Survival was difficult for the Mold at first. There were only few abandoned chips of wood on which to live, and it was nearly impossible to eke out an existence on such a small quantity of food. Not only that, but the wood was old and dry and especially difficult to dissolve with the few molecules of enzyme remaining in the Mold. Yet it managed to keep alive.

After the piece of wood was ingested, and there was no more food nearby, the once resilient strand began to wane and die. But even so primitive and simple a form of life has a will to live – an unknowing, blind will, but a will nonetheless: an almost tangible force that drove the Mold to hold its own.

Although slight in mass, the Mold conserved its energy and diverted its life force from the instinctive process of expansion, to spreading its feeble body in pinpoint directions, in a last-ditch quest for nourishment.

A molecule-thin extrusion was fortunate enough to locate a long-dead spider that had been buried when the cinder block had been laid. The Mold engulfed this morsel rapidly, and with the added energy fanned out even more. The Mold continued to survive on minute organic tidbits that it found inside the cinder bock.

Mortar had been slopped at the bottom of the cinder block, filling the hole and covering the dirt. But as the mortar dried, minute cracks appeared in the otherwise corrugated surface. The Mold extended an exploratory filament into one of these narrow cracks.

The soil beneath the foundation was teeming with organic matter, both living and dead. The Mold could consume dead tissue as well as living tissue. It added earthworms and burrowing bugs to its diet of grass roots and organic detritus. Soon it began to grow exponentially.

After spreading sideways underneath the foundation, the Mold sent tendrils upward through the holes in the offset cinderblocks. It ate through the stringers and entered the space between the wallboards. This time, however, instead of ingesting everything it encountered and increasing the size of its central mass, it diversified its form by sending filaments along edible surfaces, and only tasting its newfound source of food: tabulating, so to speak, the amount of food that was available.

Slender strands bored into the studs, but did not dissolve the wood in its entirety.

The Mold grew cautiously. It expanded – paper thin and light in weight but extremely large in volume. It spread throughout the walls, then into the ceiling. Not until it knew the extent of its domain would it consume the food that lay within its clutches.

The Mold's new imperative was concealment.

Chapter 13

"That there's a rip-roaring story," Edward commented, after the professor had lapsed into silence upon completing his doleful narrative. "I wouldn't ov believed it if it hadn't come outa yer own mouth."

"I don't blame you," I offered. "Only I never had the time to disbelieve it. It all happened so fast . . . " I trailed off, not knowing what more to say, but feeling pretty confident that I got my point across.

"It's fantastic," muttered Albert, shaking his head. "A simple thing like mold. Who'd a thought?"

"I wish it _were_ fantastic," replied the professor.

I don't think that either one knew what the other was talking about, but they seemed to regard it all right, so I did not intervene and try to straighten things out.

Arlene shuddered, and said with a creaky voice, "I've never been more frightened in my life. It was terrifying the way it grew and ate. It made my stomach turn."

"You? Afraid?" I said revengefully.

"Well, after all, I'm only a woman." It happens every time. Women strive to act like men, and emulate them to a tee. Then when an embarrassing situation occurs that they can't handle, they rely on the fact that they are _only_ women. It seems to me that they want the comforts and benefits of being a man, but none of the responsibilities.

I should have said, "That isn't what you told me yesterday." But I remembered the professor's words of wisdom – and he was looking at me suspiciously – so I didn't. I let it slide.

"I suppose we were lucky to suffer only one casualty," stated the professor, after a lull in that conversation.

"Right on," said Charles.

"That poor boy." Again Arlene shuddered – she's

been doing a lot of that lately.

Albert fidgeted with his coffee cup. "You never mentioned them by name."

"Yeah. Who were the tikes?" asked Edward.

Charles and the professor exchanged worried looks, but neither offered to say a word. I thought that was odd.

I blurted, "Shane and Albert." I didn't know their surnames, and probably couldn't have pronounced them correctly if I did.

Albert Enders was aghast. His beardless face assumed a chalky pall as the blood drained out of it. For a moment he was speechless. Then he mumbled, "Not my nephew and namesake?"

The professor's nod was almost imperceptible.

Tears welled in the corners of Albert's eyes.

Edward was slightly more stoic, but only slightly.

An uncomfortable silence reigned for several minutes. There was no clattering of cups or utensils. The loudest sound was the faint breeze soughing through the screens of the open windows. The emotional impact of the situation was pregnant.

Finally Albert regained partial control of his feelings. In a cracked voice, he uttered, "Little Albert was named after me. His ma was my wife's sister. His – his – his fokes was carried away by sickness when he was very young. Me an' the missus would a' tooken him in, 'cause we ain't got no kids ov our own, but Shane's mother insisted on raisin' the little feller an' keepin' him in the village with the rest ov his people."

Albert pushed his chair back, stood up, and left the room. Through a window I saw him wandering about outside. He trod a meandering course toward the village.

The professor breathed a very deep sigh. "The death will have to be reported to the authorities in Alice Springs."

"I can do that little thing," Edward offered, staring into the dregs of his coal-black coffee. "We're goin' to the Springs to pick up supplies for us an' ammo fer some o' the other ranchers before goin' back to our spread. I'll tell Constable Harris."

"Thank you, Edward."

"Tellin' our wimmen foke will be the hard part."

"I can imagine."

After another long lull in the conversation, Edward changed the subject. "Well, Prof, what're you gonna do now? Your lab is completely useless. Far as I could see, there's not one good piece ov equipment in it.

I had meant to ask the same question, but was waiting for a more opportune time when the professor didn't appear to be so depressed.

"At this point I don't really know. Of course, I still have my duty as an educator. I suppose that the only thing that I can do is to turn in a full report and ask for more appropriations so I can purchase new equipment."

"You tell the full truth and they might question your sanity," Charles retorted.

The professor shrugged. "I turned in some fairly fabulous reports during my time at Woomera."

"Not like this."

"No. Not like this."

"What exactly did you do at Woomera?" I wanted to know.

"Due to the Official Secrets Act, I am not at liberty to discuss my participation in experiments that were conducted at Woomera."

"I know what went on there," Charles confided. "And so does everyone else around here."

"I can neither confirm nor deny any allegations that you may make about experiments that were conducted at Woomera."

I remembered what Edward and Albert – or Albert and Edward – mentioned earlier. "You know, I've been meaning to ask – "

I was about to pursue a new line of questioning about Woomera when Arlene interrupted. "But professor, they'll have to believe you when my article is published."

I didn't like her stealing my thunder, but I couldn't help but agree. "Sure. People believe everything they read no matter how fantastic or absurd it is. That's the power of the printed word."

"Right on," Charles inserted.

"And whereas Miss Hawkins' article focuses on your relationship with the Aborigines, mine will deal strictly with your scientific endeavors. I'm sure my editor will be more than happy to give you some free publicity and words of praise when he sees *my* article. Millions of people read our magazine, and I'll bet they'll eat up your science experiments with gusto. Public opinion – "

"And even more people read *my* magazine," revenged Arlene. "*My* article will give you plenty of supporters. The government will simply have to let you continue your work. It's vital – simply vital – what you do out here."

Professor Wayward waved his hands to pipe down our enthusiasm. "I thank you heartily for your votes of confidence and support, but I do not believe that the Australian government will be swayed by propaganda that is published in American magazines. Neither of your magazines is circulated here."

"But – "

"But – "

"Not only that, I am going to have to explain to the authorities how I let a pair of Aborigine lads conduct an experiment that resulted in the death of one of them. My work here has never been looked upon as dangerous. Now I have to account not only for the destruction of my laboratory, but for the demise of one of my charges."

"But it wasn't *your* experiment – " I started.

"You can't be held responsible – " started Arlene.

"I did not mean to imply that I might be found culpable in a legal sense. But these are – my people. My wards. Shane and Albert are – or were – my prize pupils. I – I do not . . . " The professor shook his head with dejection and despair.

I realized that I had quite lost sight of the humanistic aspect of the situation. My scientific zeal temporarily outweighed my conscience.

Another uncomfortable silence ensued as we all delved into our private thoughts and emotions.

Finally, Ed pushed back his chair. "Well, Prof, I'm afraid we gotta be on our way. A herd o' dingoes been harassin' the cattle, and we plumb run outa ammo to keep the numbers down."

"Of course, Edward. I quite understand." The professor stood. "You will let Constable Harris know what occurred here?"

"I'll do that little thing. Charles, you got extra ammo at your place, in case the store ain't got enough?"

Charles nodded. "In the back shed. You know where the key is."

"I 'preciate it. We'll keep a count o' what we owe you."

"You know the bridge is out?"

"What? Again? They know that crick always floods out the bridge. Why don't they fix it right?"

Charles shrugged.

"Guess we'll take the ford."

The moment was too tense, so I passed on commenting about Fords and Chevys. I pulled Arlene aside as we left the house. "Here's your chance to go to town and beat me to the punch."

"I can't leave now. Why don't *you* go?"

"With a two-day head start on me, you should have enough material for your human interest article. I've got to hang around because of this new science twist."

"I've got to interview Shane's parents. I'll go back with you and Charles."

"Who said you were invited?"

Arlene huffed, and placed her hands on her hips. "You aren't going to leave me stranded here?" She pinched her pretty green eyes. "Are you?"

"Depends."

"On what?"

"On a lot of things."

"Such as?"

"Well, I'm paying Charles by the day for his time, and until I submit an expense report, have it authorized, and get reimbursed, I'm paying him with my own money."

"Okay, I'm willing – "

"Then there's the matter of you being my competitor."

"But I thought we settled – "

"And, of course, there are specific instructions from my boss not to fraternize with the enemy."

"But I'm not – "

"And finally, there's professional courtesy."

"I've been courteous – "

"I mean in a professional manner. I don't remember you offering to share your bush plane with me in Darwin."

"Would you have wanted to parachute – "

"That's not the point. The point is that you didn't make the offer. You took off on your own so you could steal a march on me. You were in such a hurry to get here that you didn't make arrangements to get back. Now you want me to make up for your lack of foresight."

"Well . . . "

I glared at her hard.

"But you're not me. You're you. I might do something like that, but you wouldn't."

"Are you sure?"

She glared back at me.

"You've got a bed to sleep in. I've got a tent and sleeping bag."

She almost choked on her sudden intake of breath. "You aren't suggesting . . . "

I grinned and walked away.

"Well, I never!"

Over my shoulder, "And probably never will."

She had no riposte for my parting shot.

Albert and Edward – or Edward and Albert – were already sitting in the cab by the time I reached the pickup truck. Albert was sulking in the passenger seat.

Shane's mother – Little Albert's foster mother – was standing outside the cab with her arms through the open window and wrapped around Albert's neck. She was crying softly.

Edward grabbed the steering wheel and turned on the engine. Little Albert's foster mother disengaged herself from Albert. Albert stared sightlessly through the dusty windshield.

Charles motioned westward. "Looks like dark clouds heading our wiy."

Edward glanced out the window. "Storm's a brewin'. We better hurry if we 'spect to cross the ford."

Charles explained to me. "The ford becomes impassable after a downpour."

"For how long?"

"Couple, three diys. Maybe longer."

Edward threw in the clutch. He waved as the truck bounced along the dirt path that served as a road. "Adios, amigos."

Now it seemed as if I might be stuck here with Arlene for the better part of a week. I toyed with the idea of asking Charles to take me back to Alice Springs right away, before Noname Creek flooded Noplace Valley. I abandoned the idea almost at once. I was too old a hand at this game to give up prematurely with half a story. I decided to stick it out, learn more about the professor's erstwhile experiments with herbs, decipher what I could about the newfound Mold, and delve into the secrecy and intrigue that surrounded the professor's previous port of call: Woomera.

Come hell or high water, literally, there was nothing for it but to meet the approaching storm head on and sit out the rising tide.

Chapter 14

Rain fell hard all night long. I remembered my Boy Scout training, and dug a trench around my tent so the water could drain away. I also dug a channel to drain the trench into a shallow depression.

I hung a flashlight from a string that I tied to the pole so I could see to write in my notebook. I wore down my pencil point and sharpened it twelve times with my penknife before my eyelids got so heavy that I could no longer keep them open.

The sky overhead was still overcast in the morning, but there was a glimmer of sunlight where the clouds were clearing to the west. The night chill clung without warming sunbeams, so I stayed wrapped in my sleeping bag long after I heard the buzz of activity from the nearby village. Soon, however, I was going to have to face the raw air and beat a retreat to the outhouse.

The soft patter of dainty feet stopped outside my tent. "Knock knock."

"Arlene?"

"None other."

"Come on in and make yourself at home."

She unbuttoned the flaps and flung one back. She was sitting on her haunches, dressed in boots and outback khakis. She was hatless, and her red hair was combed and flowing smoothly over her shoulders. She was smiling. "Sleep well?"

I stayed snuggled in the sleeping bag. "Did you come here just to rub salt in my wounds?"

"No, actually, I came to make a proposal."

"I don't accept proposals before breakfast or marriages before lunch. What's on your mind?"

"Well, all night long I couldn't help but think of you out here in this tent. In the rain. In the cold."

"That makes us even. I kept thinking of you in your warm dry bed, with sheets, blankets, and an emu down comforter."

"The comforter is stuffed with cassowary feathers."

"Whatever."

"So I thought we could compromise and share the bed."

This made me sit up and stare wild-eyed. I rubbed the goosebumps on my bare forearms. "Well, I never – "

"We'll take turns. I'll sleep in it one night, and you sleep in it the next."

This wasn't exactly the proposal that I wanted to hear. "Well, it's a pretty big bed. We could sleep on opposite sides, or maybe – "

She blushed cherry red and her freckles popped out. "Don't even think about it, Mister." She threw down the flap and stalked back to the house. I guess that was equivalent to slamming a door in my face.

If you don't ask, you'll never know. But I already knew. That woman could be so exasperating at times. All the time. Well, most of the time.

"Hey, Mite. What got into the lidy? One second she was smiling, the next she was scowling."

"I know. She's a quick change artist when it comes to moods."

"The fire in her eyes could have ignited wet logs."

I threw back the flap. "One of us made an improper proposal."

Charles looked down at me with arms akimbo. "Which one?"

"I don't know." I noticed that his clothes were clean and dry, and there were no footprints in front of his tent. "Hey, where'd you sleep last night?"

"In a hut."

"Oh." Then I thought about it some more, and realized the implication. "*Oh!*"

Charles winked. "Watch yourself. That's one strong-minded woman."

"Don't I know it?"

"Abo women aren't like that."

"You mean fickle?"

Charles grinned and nodded.

"That must be the reason Edward and Albert – or

Albert and Edward – took Abo wives."

"Right." Charles kept grinning. "Let's go get some chow."

"I'm with you there."

"Ciao."

I dressed quickly, rushed through my ablutions, and joined the group in the dining room. The warm, dry dining room.

Arlene was running hot and cold: she gave me a brimstone stare and a cold shoulder at the same time. "How long are you going to suspend classes, Professor?"

The professor nodded at Charles and me as we took our seats. A pot of hot tea sat on the table, and cups were set on placemats in front of the chairs. Marie, cheerful but quiet as always, served fried blood and gruel. I ate the gruel.

"I may as well resume school today, and work on the laboratory after hours. I will not assign homework. The boys and girls will gladly help to straighten out the laboratory."

"I don't think it can be strightened, Doc."

"Let us say, emptied of debris."

"How is Shane holding up," I asked.

"Not well, I am afraid." The professor sipped some tea and took a large bite of black blood. "Last night he was uncommunicative. This morning he is catatonic."

"Sounds like a bad case of shock."

"That is my diagnosis."

"After all, he *did* see his friend eaten alive."

I didn't mean to sound heartless, but judging by the daggers that darted out of Arlene's eyes, she must have interpreted my statement that way. Or maybe she was just looking for an excuse to knife me.

"True, true. That must have been quite a traumatic experience for him. For all of us, really. But children are not equipped to deal with trauma the way adults are. I can only hope that he snaps out of it before he starts to deteriorate physically."

"Maybe I can help."

"I was hoping that you could, Charles."

Charles explained to Arlene and me. "Part of Shine's problem stems from his parents. They're from the old school. Actually, no school. No progressive school, that is. They still harbor Aboriginal beliefs and superstitions. Shine's parents let him attend Doc Warren's school, but they don't go as far as allowing modern medical treatment."

"Aboriginal cultural development is still in a transitional phase," added the professor.

Arlene nodded in agreement. "Social science teaches us that all cultures exist in a constant mode of transition. There is always a dichotomy between tradition and progress; between the status quo and the state of change. Cultures must evolve in order to survive. Any culture that remains static eventually ceases to exist.

"It is also true that no culture is ever truly homogenous. Every culture contains some subdivisions that cling to the past, some that abide in the currently accepted framework, and some that live on the fringe and possess the elements of change. In the U.S., for example, we have American Indians who want to live in an imaginary past that no longer exists; members of the Church of Christ who decline medical treatment and pray for cures; beatniks who dance to the beat of a different drum; and splinter groups of all kinds who profess or believe in any number of wild and illogical ideas. And while the majority of people may follow the mainstream, there is a strong multicultural component – in any society."

If I hadn't been angry with her, I would have applauded her perspicacity, or at least said, "Here, here." On the same hand, I could have faulted her oxymoronic statement about "an imaginary past that no longer exists." By way of appeasement, I kept my own counsel in both regards.

"It appears, then, Miss Hawkins, that the social imperative parallels biological evolution. For better or for worse, both contain the seeds of change."

Charles said, "Shine's parents are treating him with

local remedies. Now, I have a great deal of faith in local remedies because I was brought up on them. As Doc Warren will tell you, there are herbal medicines in Austrilia that are far more effective in the cure of disease than many high-priced pharmaceuticals. But as *I* can tell you, some of these herbal medicines are placebos. Thiy are nontherapeutic. If thiy work at all, thiy work because the pitient believes thiy will work."

"In other words," added the professor. "They can cure many purely psychosomatic disorders, but are powerless to treat the symptoms of organic or functional diseases."

I raised my hand like a student. "But if Shane is suffering from a traumatic experience, isn't that a psychological condition? One that medicine can't cure? And that psychiatry can?"

"Ordinarily that is true, but not in Shane's case. His parents permitted me to conduct a brief examination this morning. His breathing is shallow and stressed. His pulse is faint. His heartbeat is irregular. His pupils are restricted. His muscles twitch. I am not a doctor, but I can tell you that these are not the normal symptoms of shell shock or battle fatigue, which I had occasion to treat when affected soldiers were removed from the field and reassigned to Woomera – a place that was relatively quiet during the war. Overstressed pilots regained their confidence by practicing aircraft takeoffs and landings on a mock runway. I assisted the base doctor in performing their medical workups." In aside, "That information is no longer classified."

Arlene brightened considerably. This was precisely the kind of information that she needed for the focus of her article. Whenever she brightened, I brightened, because she did not look upon me so darkly.

I returned to my tent after breakfast and did some work on my article. Its length had already exceeded the planned space limitations, but I found a way to make it a two-part piece by ending the first part with the professor's past experiments, and beginning the second part with the introduction of the Mold. Old Skinflint would love this, because he would get two articles for the price of one – perhaps two for the price of none if he failed to reimburse me for my expenses.

The sun was making an attempt to peer through the overcast, but wasn't having much success. Either the storm had stalled, or another one was following close on its heels. I sprawled out with my legs and lower abdomen inside the sleeping bag, lying on my left side so I could write with my right hand. A sweater warmed my upper body. I wasn't toasty, but I was passably snug.

I viewed approaching lunchtime with mixed feelings: I was hungry, but I wanted to stay out of Arlene's line of fire. That was when I noticed the artillery coming into range. The expression on her face was inscrutable. I geared myself for the worst.

"Knock, knock."

"If this is another knock-knock joke, I'm not home."

"My, we're touchy." Arlene squatted on her haunches in front of the tent.

"Look who's talking. I mean, you look at me in a friendly tone of voice, I make an innocent comment, and you run off in a huff with steam boiling out of your ears. Does the word wishy-washy mean anything to you?"

"Steam doesn't boil. And I told you already, wishy-washy is two words."

"Whatever." I felt as if I were on a merry-go-round. "So what face are you wearing this time? The Good Witch of the North or the Wicked Witch of the West."

"Stop being cynical."

"You give new meaning to multiple personality disorder. The three faces of Eve can't compare to the menagerie that's running around in your mind."

"Tim!"

"What?"

"I'm trying to apologize for this morning. Stop making it difficult for me."

I was momentarily nonplussed. I put down my pad and pencil and sat up in my sleeping bag. "You know, if you just maintained an even keel, you wouldn't have to apologize so much."

"I know. I know. And honestly, I'm working on it. It's just that every male reporter I've been associated with has belittled me because I am a woman. Either because they didn't think I was competent enough to do the job, or because I was a threat to their masculinity."

"We've already been down this road. I never – "

"I know you didn't. You made your jokes and innuendoes, and I construed them in the worst possible way. You were trying to be funny. I was trying to be stern and aloof. That's just my job hat – "

"Or face."

" – but that's not who I really am. I thought if I acted tough I could meet you on your own terms. Now I realize that you don't act tough. You don't have to act tough. All you have to do is act natural, er, naturally."

"Sounds like last year's number one song on the Hit Parade."

"See what I mean. You're always joking."

"Life is too short to be serious all the time. People need to lighten up some."

"I know. I know. I realize that now. You're not like the other reporters. You're – different. And I want us to be friends."

"For how long?"

She tilted her head. "For as long as you can stand me."

I glanced at my watch. "That didn't last very long."

"Oh, you."

"Arlene – "

"Tim, there's one other thing."

"Uh-oh. I knew it was too good to be true."

"I'm serious."

"So is the Dog Star."

"Huhn?"

"Sirius is the brightest star in the sky. In the constellation Canis Major."

"Sorry, but I don't know science stuff. I didn't understand half of what Professor Wayward was talking about yesterday. Fungus and slime molds and – "

Arlene burst into tears with the abruptness of a rainsquall. I suppose she must have been holding it in since yesterday. She threw up her hands, lost her balance, and fell forward on top of me. Her weight pushed me backwards. We sprawled across the sleeping bag like an amorous couple in heat, except that she was bawling to beat the band. That's bawling with a w and not a double l.

I wasn't used to having sobbing women on my hands, or on my body either. I didn't know what to do, so I just let her have her cry. She went at it for five minutes or more. I couldn't see my watch. She wound down slowly, like a child's top that wobbled as it lost its spin.

"I'm – I'm sorry – Tim. I didn't mean – to cry all over you."

"You shed enough salty tears to kill two Molds."

It was the wrong thing to say, and the wrong time to say it. She burst into tears all over again. Another five minutes passed before she wound down again, like a siren receding in the distance. I would have enjoyed the pressure of her body against mine if the situation hadn't been so grave. Actually, I enjoyed it anyway, particularly as she made no attempt to get off of me.

Arlene lay limp across my chest. "I'm sorry, Tim. I didn't mean to throw myself at you like that." But neither did she make any attempt to remove herself. Instead, she nestled her head between my pectoral and armpit, and wrapped her arms around me.

"How did you mean to throw yourself at me?"

She put on a pouty face. "Did I crush you?"

I could hardly breathe, but I daren't let her know. "You're as light as a feather and as sharp as a quill."

"There you go again."

"It's my nature."

She wiped tears off her cheeks with her shirtsleeves. "It's a nice nature. Don't lose it. Especially on account of me." She shifted her weight with a grunt. "Is that a pencil in your pocket?"

I avoided the obvious Mae West rejoinder like the

plague. I luxuriated in the full-body touch of Arlene's womanhood. "Does this mean that you no longer think of me as an old reprobate?"

"You're not old."

"And that doesn't answer – "

I was cut off by a resounding crash that shook the very earth. It sounded like an entire forest had dropped out of the sky, as if some titanic Cyclops had decided to play Pick-up Sticks with redwoods.

I gasped, "What the hell – "

But it was worse.

Chapter 15

I tossed Arlene aside like a sack of potatoes.

When I sat up and peered out of the tent, I saw an astonishing sight: the schoolhouse was literally palpitating. The walls were expanding and contracting like a bellows. My first impression was that the building was alive and breathing. But then I noticed that it was also shrinking, like a balloon that was deflating in extreme slow motion.

"Holy smokes."

Arlene peered past my shoulder. "Tim – "

"I don't know." I crawled out of the tent lickety-split. I tossed aside the notepaper that Arlene had crumpled on my lap. I said it again, "What the hell – "

The schoolhouse walls were no longer vertical or perpendicular to each other. Planks splintered loudly as I watched in wonder and awe. You can't imagine the horrific noise that a building makes when it is being crushed and sucked into itself. Adding to the cacophony were the screams and shouts of children. The building now had the appearance of a brown paper grocery bag whose sides were crinkled and pinched in the middle.

Arlene gasped and cried something indiscernible.

The schoolhouse was collapsing upon itself.

While it may seem that I was frozen with fear, and dazed by the mind-boggling sight, it must be understood that the apparent slow-motion destruction of the schoolhouse took place in only a few seconds. My temporal frame of reference was grossly exaggerated.

Children erupted from the schoolhouse doorway with wails and caterwauls that sounded like the sirens of a dozen fire engines and hook-and-ladder trucks. Shivers coursed along my spine as if a teacher had just scratched chalk along a blackboard.

I was galvanized into action. I started running for the schoolhouse. I didn't know what I was going to do;

I just felt that I needed to be closer to the source of trouble.

Windowpanes cracked and splattered glass like shrapnel. Frames and casements distorted as the wood bowed then splintered. The clapboard façade assumed impossible angles. I gazed upon a building that could have been imagined only by someone like Salvador Dali.

The tinkling of glass merged with the snapping of boards and the yells of children as they scattered in all directions. Professor Wayward stumbled headlong onto the soft earth. A small child he was holding banged hard against the ground and rolled over. The tot was so startled by the precipitous bump that it did not even cry. The professor's arms and legs moved ineffectually. I skidded to his side on my knees, like a baseball player sliding into home base.

"Professor!"

The professor spat dirt. "Four more – children – inside. Far corner. Must – get them – out."

I looked up at the schoolhouse. There was no way to get inside because there *was* no inside. The doorway was constricted like a closed diaphragm, and the building itself was nearly flattened: not only from the top but from all four sides. Whatever construction materials hadn't yet been pulverized were blanketed with a white fluffy substance that reminded me of cotton candy.

The entire schoolhouse lay engulfed in Mold!

One crazily canted wall threatened to topple on top of us.

"Tim!" Arlene scooped up the child and backed away.

I grabbed the professor by one arm and commenced to drag him from the path of danger. He was heavy but not overweight. I was not making much progress until Charles stooped on the professor's other side and took a firm grip under his armpit.

"Together now, Mite."

Charles was much stronger than I was. We lugged the professor along the ground like a sack of potatoes.

Despite the terror of the moment, part of my mind recalled that I had just used that simile in my mind. We lugged anyway. We laid him on his back a safe distance from the schoolhouse and encroaching Mold.

The professor coughed. "The Mold." Cough. "It broke loose." Cough. "Came right out of the walls." Cough. "All at once." He needed a drink of water but I had none to give him.

What remained of the schoolhouse was a horrible pulsating mass of downy white. The sound the Mold made as it ingested every bit of organic matter reminded me of some prehistoric monster as it masticated a forest. Already I saw patches of Mold extending toward us from the central mass, as if it could sense our presence.

"What can we do?" I shouted. "We're out of salt."

"Burn it," Arlene replied. "Incinerate it. Cremate it."

Charles grinned like the Cheshire cat. "Brilliant."

"Will it burn?" I wondered.

"It will if we soak it with petrol." Charles jerked his head in the direction of the generator shed.

I noticed that the generator was not making any sound. It must have short-circuited.

"Do it," coughed the Professor. "Use the petrol for the generator. Use all of it."

By this time a crowd of villagers had gathered behind us. I saw several women crying, beseeching with arms held out for their missing children. I stifled a momentary feeling of compassion in order to concentrate on the task at hand.

"Let's go," I shouted, but Charles was already on his feet and running ahead of me.

I caught a glimpse of an Aborigine mother taking the tot out of Arlene's arms. Then I turned my attention to the shed. Charles fumbled a key out of his pants pocket, released the lock, and flung open the door. No words were needed. He grabbed two jerry cans; I grabbed two more. Each can held five gallons of gasoline, which weighed nearly forty pounds plus the weight of the cans.

The cans bumped and grated awkwardly against my legs. I struggled and straggled behind Charles, dragging the cans across the dirt. By the time I reached the grounds near the schoolhouse, Charles already had his clasp knife out. He unfolded the blade and used it as a pry bar to twist off the cap by locking the blade across two opposing ears.

"We need some way to spread the petrol," Charles announced.

"I'll get some pots and pans from the kitchen." Arlene was off like a flash.

Charles shouted in Aborigine dialect, and half a dozen women ran after Arlene. By the time they returned with armfuls of cookware, he had the Aborigine men lined up like British soldiers making a frontal assault. Each man took a pot or pan from the women, and held it under the spout while Charles poured gasoline into the container. I used the knife on another jerry can.

"Don't get too close," cautioned the professor.

Aborigines surrounded the schoolhouse like a tribe of redskins attacking a wagon train. Each one doused the perimeter of Mold with the petrol from his cookware, then dashed back to Charles for a refill. In a few minutes the air reeked like a gasoline spill after a highway crash.

"Now all we need is a source of ignition," I commented dryly.

The professor calmly pulled a pack of matches out of his smock pocket. "I use them for lighting Bunsen burners."

At the moment I didn't care if he was a closet pyromaniac, as long as he was going to torch the Mold before it escaped. Already the Mold was on the move from its roots. First tendrils, then fingers, then armlike protuberances spread outward, like an amoeba on the prowl. The Mold grew and expanded as it consumed the schoolhouse and the bodies that were trapped inside.

Charles tilted a jerry can and laid a trail of gasoline from where the professor still perched on the ground, to where an extension of the Mold pointed directly at us. After Charles signaled okay with a single nod, the professor scraped a match across the striker. The match burst into flame.

The professor stared at the match, glanced at the Mold, then without another thought let the match drop into the pool of gasoline. I felt a gut-wrenching twinge when the match sputtered out in the liquid fuel – seemingly drowned before it could ignite the fumes. Then a barely perceptible pinpoint of light flared abruptly into existence. Flames shot a foot into the air, and raced along the octane pathway like black powder sparks in a Western movie.

Charles let loose with a deep-throated cheer.

Like the scientist that he was, the professor retorted softly, "Let us see if the Mold can consume hydrocarbon as readily as it consumes carbohydrates."

A ghastly thought haunted the back of my mind. "Professor, aren't there some molds that can grow on petroleum products?"

"No. Those are bacteria. Bacteriologists are experimenting with petroleum reducing bacteria as a way to contain oil spills from stranded tankers. The fermentation process – "

The professor's words were drowned out by the explosion that was caused by the sheet of flame that erupted in a circle around the Mold. Fire roared and crackled with abandon. A long, drawn out squeal made me cringe in fear, because my first thought was that the Mold was screaming in agony. But the noise was made by the release of air from pockets in the wood. I kept cringing nonetheless.

Aborigines brought five more jerry cans from the shed. Charles ladled out petrol in generous portions. The men approached as close as they dared to the blaze, tossed their liquid fuel into the fiery cauldron, then retreated from the excessive blast of heat. The Mold was burning from the outside in, but the center appeared to be untouched by fire.

"Is the Mold itself burning, or is it only the wood

that the Mold hasn't consumed?"

The professor studied the conflagration with scientific detachment. "An excellent question, Mr. Baker. I have never tried to burn mold."

"It's ironic that yesterday we fought the Mold with fire extinguishers, and today we're fighting it with fire."

Arlene was always thinking two thoughts ahead. "Can't we squirt some gasoline into the center somehow?"

"The only pump we have is the water pump in the well, but that would not do the job even if the generator were working."

"How about Molotov cocktails," Charles suggested.

All three of us looked at him questioningly.

"I've got a bunch of canteens in the Land Rover. How about we pour out the water and fill them with petrol?"

Arlene was hot on the topic. "I saw some plastic water bottles in the kitchen."

To say it was to do it. Charles rallied the Aborigines to the task.

"There is a funnel for the generator in the shed," observed the professor.

I retrieved the funnel. In two shakes of a lamb's tail, Charles and I were pouring gasoline into containers, and Aborigines were hurling them into the center of the erstwhile schoolhouse. The water bottles worked best because the plastic melted from the intense heat, and exploded. The canteens were made of aluminum; the thick screw caps melted slowly, and when they did, the gasoline had to either pour out of the canteen to feed the fire, or, if the canteen landed upright, the fumes burned upward but did not douse the Mold. Liquid gasoline doesn't burn; only the fumes burn, by admixture with oxygen in the proper amount.

The bonfire raged.

"Did you see any marshmallows in the kitchen, Arlene?"

She gave me a look that could have turned Medusa to stone.

"Guess not," I muttered, in a halfhearted attempt to save grace.

The burning school was the Mold's funeral pyre. Flames shot twenty feet into the air. Aborigines shuffled or cavorted around the fire, in a native rendition of a circle dance. They tossed the last dregs of gasoline wherever they saw a speck of Mold that hadn't been totally consumed. We never did resolve the issue of whether the Mold actually burned, or whether only the gasoline burned and it was the heat that destroyed the Mold. I suppose the point was moot as long as the Mold was killed.

"Stubborn stuff," I commented.

Professor Wayward nodded. "I speculate that some of the Mold escaped our notice by growing inside the walls, where the ionized water could not reach it. What puzzles me is what possessed it to remain completely hidden until its explosive breakout. It is almost as if it knew . . . "

I started to get tingles from the dire implications. "How can Mold 'know' anything? I mean, it's just mold."

The professor stood in quiet contemplation, squeezing his beard constantly and kneading his face. After several minutes, he said, "There is a great deal that we have yet to learn about molds. About all fungi. Some taxonomists believe that certain species of what we call fungi should be placed in separate kingdoms. Life at the unicellular level is difficult to categorize. Even multicellular organisms are difficult to categorize. Taxonomy exists in a constant state of flux."

I remembered what he had said about this Mold being an incipient organism. Mentally I recited the taxonomical classifications in descending order: kingdom, phylum, class, order, family, genus, species, variety, type. Mankind was placed in the animal kingdom, along with salmon, jellyfish, turtles, bugs, and brontosauruses. In binomial nomenclature, mankind belonged to the genus *Homo* (man in Latin) and species *sapiens* (wise): *Homo sapiens* meant literally "wise man" (although sometimes I found exceptions to that specific designation, especially in politicians).

If I inferred correctly from the professor's working hypothesis, this Mold might not represent just a new species, or a new genus, or new family, but a new phylum or kingdom.

"Could this Mold be a mutation?" I asked.

"I do not know. Part of me wishes that I could retain a sample for study. But another part – a saner and more rationalistic part – knows that we must treat this organism as a contagion. We have seen how quickly it can metastasize. Imagine if we were to accidentally release it upon the world. It would be devastating. It possesses the potential to consume all life on the planet."

"The end of the world as we know it," I said.

"So we are fighting Armageddon," Charles prophesied. "The final battle between good and evil."

The professor's eyes were steely. "Good and evil do not exist in nature; only winners and losers."

"And the fittest survive," I declared.

"And the fittest survive," repeated the professor. "Pray that we are the fittest."

Chapter 16

The fire was a long way from dying out.

The outer edges were reduced to smoldering cinders, but the core of the schoolhouse still burned furiously. A plume of black smoke rose hundreds of feet into the air. The fresh fragrance of burning wood contrasted sharply with the caustic odor of melting insulation. I could see bare copper wires among the ash.

I watched the flames with considerable apprehension. The cleansing fire crackled and popped discordantly. I was mesmerized by the flickering red tongues that danced sinuously with the vagaries of the breeze that ruffled my hair.

Marie crouched next to the professor. She had brought a chair from the house so he could sit comfortably in front of the holocaust. She had also brought jugs of water from the emergency supply, stored against times when the electric well pump ceased to operate. She poured water into earthenware mugs which she dispensed to a constant stream of thirsty people. My throat was still parched despite two mugs of tepid water and an overcast sky that reduced ambient temperature. After Marie ran out of stored water, the Aborigines retreated to the well nearer the village – the well that was operated by a hand pump. They refilled the containers.

Professor Wayward watched wistfully as both his dream and his nightmare went up in smoke.

In the distance I saw a small herd of kangaroos. A few joeys that were fresh out of the pouch bounded playfully, but the adults stood stock-still and watched the fiery spectacle intently. Far off, to the north and south, the MacDonnell Mountains stood in stark relief against a largely flat and featureless plain.

For a moment I felt that the land had returned to that immeasurably long period of placidity before man and Mold had invaded this primitive and isolated subcontinent. But only for a moment.

A terrifying ululation drew everyone's attention away from the conflagration. A young female Aborigine was running toward us from the village, screaming hysterically.

"Now what?" I murmured under my breath.

"Shane's mother," uttered the professor.

The woman threw her arms around an Aborigine male whom I assumed to be Shane's father. She jabbered so fast in local dialect that the professor could not understand her.

Even Charles understood only a few words. "Something's happened to Shine, but I can't catch what. Something about . . . " He shook his head. " . . . hair." He shrugged. "Shine's hair."

I needed no further elucidation. Whatever had happened to Shane, it must be bad. Charles and I exchanged brief looks, then we took off running. I beat him to the hut. The building was more of a cabin than a hut, framed with studs covered with clapboard instead of crudely made from logs or rough-hewn timber.

I burst through the outer door into a communal room that had a cooking area at one end. Three doorways in the rear wall led to bedrooms; one door was open. I charged through the opening, but came to a screeching halt by placing my hands firmly against the jambs. I couldn't believe my eyes.

Charles stopped behind me, breathing hard. "What the bloody hell?"

Shane's body reposed on the bottom platform of a bunk bed. He wore nothing but a pair of black shorts. Snow-white hair as fine as fluff sprouted from every pore. It reminded me of a downy chick, or a dandelion that had gone to seed. The skin was shriveled and the abdomen was oddly shrunken, as if ancient Egyptian embalmers had removed the organs and placed them in canopic jars. An image of a kewpie doll passed through my mind.

"It's Mold," I whispered, barely audible to myself let

alone to Charles.

"Growing on him like that?"

"Not growing *on* him – growing *inside* him." I heard feet shuffling across the floorboards. I turned and saw Arlene and Professor Wayward, their faces a mixture of fear and curiosity. "You'd better look at this, Professor."

Charles stepped aside so the professor could see through the doorway. "Oh dear."

"The Mold is consuming him from the inside."

"I see that." His words were soft-spoken, but his tone was one of repugnance. "We must – we must destroy the corpse."

"Right on," said Charles.

"Not only the corpse, but the room. The building." We all stared at Arlene as if she were crazy. "Everything that boy touched could be infected with – Mold. And everyone who touched him could be infected, too."

"She's right, Professor. You yourself said that a single strand might be able to produce a new individual."

"So I did."

I could see the scientific wheels turning inside the professor's head. "There is ample evidence in the record of life that organisms can survive traumatic injury, and can reproduce themselves in a variety of ways."

I offered confirmation. "A starfish can grow a new arm after one has been cut off. Lobsters and crabs can regenerate claws. When a flatworm is cut in half, the head grows a new tail and the tail grows a new head."

"All quite true, Mr. Baker." Squinting hard, the professor looked at each of us in turn. "This Mold is phenomenal in its ability to survive. If it reproduces by budding – an asexual process in which cells divide to produce outgrowths that are detached from the parent organism – then perhaps not only every strand, but every cell possesses the potential to become a new individual."

"Suppose it does both? Budding *and* sporing?"

"Oh dear." The professor stared at me as if I had just slammed a two-by-four into his gut. "Oh dear."

"And suppose it can also flow across the ground like

slime mold?" suggested Charles. "Leaving tracks of pure protoplasm."

"Oh dear."

I had a worse thought. "And suppose the Mold on Shane can release spores into the air. If we inhale those spores . . . "

The professor was too shocked to oh-dear us again. He stepped into the room and bent close over the body. Shane now looked like a sheep shorn of its wool but still wearing its thick undercoat. After a moment's cogitation, he said, "Miss Hawkins is right. We must destroy this corpse and everything around it. We must take all precautions that the Mold does not spread."

"Not even a snippet for study?" I offered.

"Nothing. No buds, no spores, no cells, no naked protoplasm. No vestige must remain."

"We've already used all the gasoline for the generator. And the two extra cans for the Land Rover."

Charles was thinking fast. "There's still petrol in the tank."

I was thinking just as fast. "Can we siphon it out with a hose." I used to do that when I was a kid, to get gasoline from my father's car for the lawn mower.

"No need. I can pull the feed line off the tank."

"I'll get some basins from the kitchen."

We raced out of the hut like cats after a canary. By the time I got to the Land Rover with the basins, Charles was already lying on the ground under the chassis, sprawled between the rear wheels. I crawled between the tires on the driver's side, pushing two washbasins in front of me. With a short-handled screwdriver, Charles unscrewed the hose clamp that secured the rubber tube to the gas tank.

I positioned the basin under the inlet. "You've done this before."

"Many times, Mr. Biker." He pulled the tube off the inlet.

Gasoline spurted out of the tank into the basin. Some of it splashed off the white enamel onto my face. When the basin was three-quarters full, Charles plugged the inlet with the tube. I dragged the wash-basin out of the way and positioned the other one under the inlet. Charles pulled the tube off again. The basin filled only halfway when the flow of gasoline was reduced to a dribble. Charles replaced the tube but did not bother to replace the hose clamp.

We walked as fast as we could with the washbasins, careful not to spill a drop of the precious fluid. Charles carried the half-full basin into the bedroom. Arlene grabbed the other side of my basin and helped me ease it through the outer doorway into the common room.

Charles covered the corpse liberally with gasoline, then used what remained in the basin to douse the walls and floor. Arlene helped me to pour gasoline into the corners of the common room, and into each of the other bedrooms. We saved a little to toss onto the roof.

The professor pulled a sheaf of papers from the pocket of his smock, rolled one sheet into a cone, touched the open end with a lighted match, waited for the paper to catch fire, and tossed the cone into the hut. The paper did not land on a patch of gasoline, but it rolled in a semicircle so that the flame contacted the fumes from a nearby puddle. The floor burst into flames with a whoosh. The fire quickly spread to the corners of the room and into the adjacent bedrooms. Within minutes the hut was engulfed in flames.

Behind me I noticed a forlorn Aborigine woman: Shane's mother. Her left hand was cupped over her mouth; her right hand was gripping her abdomen, with the fingers digging into the flesh. Two tears rolled down her black cheeks from even blacker eyes. Her husband stood beside her with his arms hanging straight down, fingers loose, but with a haunted expression on his face. In the course of two days this couple had lost two boys: one son and one foster son.

"So sad." Arlene was crying too. "So very, very sad."

We had won another battle against the Mold. But I was beginning to wonder if we had yet won the war.

Chapter 17

The Aborigines were in mourning over their lost loved ones. They retreated to the village where, in lieu of burial ceremonies, they held a communal service in which everyone participated. Everyone but Marie, that is, for she remained faithful to the professor and his immediate needs. Her usual cheerfulness was gone.

Professor Wayward, Charles, Arlene, and I sat around the dining room table, drinking warm tea and eating biscuits (or as we call them in the States, crackers).

For a change I had nothing witty to say, but I was about to make an astute observation when Marie screamed like a banshee in the kitchen. This was followed by the crash of breaking crockery. Goosebumps rose on every square inch of my body, and a searing burst of heat coursed along the full length of my spine.

Charles and I leaped to our feet simultaneously.

Marie erupted from the kitchen with her arms flung up in the air. Words came out of her mouth between screams.

"Mold," Charles translated.

We stepped cautiously into the kitchen. Broken dishware lay all over the wooden floor, along with the victuals that Marie was preparing for our dinner. A large fruitcake sat on the countertop. Upon close inspection, what at first I mistook for frosting turned out to be a greenish paste of mold. My heart raced with fear.

The professor and Arlene entered the kitchen on tenterhooks. I pointed silently to the cake. I extended my finger, but kept my arm tucked tight against my side. The professor pushed between Charles and me. He never took his eyes off the cake and its dull green coating. Slowly he pulled open a drawer, withdrew a long carving knife, and stabbed at the fringe of mold.

"Humph," he commented dryly. He cut away a slab

of cake on which the mold was growing prodigiously. "Genus *Aspergillus*, but I am not certain of the species. There are more than one hundred species, you know."

Marie had stopped screaming, but I wanted to take up where she left off. She had scared the living daylights out of me, although I would never admit that in public – especially within Arlene's hearing range. It took every bit of effort and willpower to get my breathing back to normal. My body was soaked with perspiration, and I still felt intense heat from the adrenalin rush.

"Not entirely unexpected. This cake is four days old, and the relative humidity has been fairly high because of inclement weather. Some species of *Aspergillus* are particularly attracted to sugars and starches."

The professor dissected the cake and cut away the portions that were moldy.

"*Aspergillus* has short roots so we can eat the unaffected parts of the cake."

"I'll pass," I said with resolve.

Charles agreed with me. "Count me out."

Arlene cut a slice and took a healthy bite. "Tastes good to me."

Someday I'm going to figure out that girl.

Some day. But not today.

Charles and I helped Marie to clean up the kitchen. Then we took our usual places at the dining room table. She prepared more food while we discussed current events and made small talk.

"I went on a jamboree when I was a Boy Scout," I started, sipping tea that had reached room temperature. "One thing I remember is that the scoutmasters taught us how to shave wild scallions into a frying pan to mix with our ground beef. Scallions taste a lot like onions. It was like having hamburgers with onions. Anyway, that night we went for a hike in the woods, and along the way I saw a dim green glow on the soggy ground."

"Phosphorescence?" Charles suggested.

"Close but no cigar. No, it was a species of light-emitting fungus."

"Are you making this up?" Arlene asked accusingly.

"Cross my heart and hope to die." I crossed my heart symbolically. I said nothing about male deer or black hair; I had used that line already. "No, it really *was* fungus. This fungus was growing on a rotten log. I broke off a piece and held it up to examine it, but when I shone my flashlight on it, the glow disappeared, and I couldn't see anything that caused it. Then, when I turned off the flashlight, the green glow came back. The light from the flashlight was so bright that it washed out the light from the fungus. And the fungus was all but invisible in the light.

"At the time, I didn't know it was fungus that caused the glow. A scoutmaster explained it to me afterward. Anyway, the green glow was so bright in the dark that when I held the piece of log next to a newspaper, I could actually read the newsprint."

"You *are* making it up."

"Right on."

"No, really. It happened. Honest Injun."

The professor snickered. "The phenomenon is quite common in nature and takes many forms. It is not phosphorescence, however, as Charles suggested. Phosphorescence refers to the luminescent emission that occurs when the element phosphorus is irradiated, causing the phosphorus to oxidize faster than it does normally. Luminescent hands on wristwatches are coated with phosphorus so they can be seen in the dark.

"What Mr. Baker experienced was bioluminescence: literally, light from life. It is prevalent among benthic fish and other deep-water organisms. Boaters often see bioluminescence in the wake of their boat at night, as the propeller churns through tiny organisms whose tissues are torn apart, causing the release of certain chemicals that react with seawater.

"The most well-known example of bioluminescence is the firefly, or lightning bug, and the glowworm which is its larval form. In biological terms, this insect is actually a beetle and not a fly, bug, or worm. A controlled chemical reaction in the beetle's tail creates a luminescent burst that is very similar to the sustained light that Mr. Baker described.

"Dozens of species of fungi are bioluminescent. Foxfire is the word that describes what Mr. Baker observed. At one time it was believed that the luminescent phenomenon called will-o'-the-wisp or jack-o'-lantern was the result of the spontaneous combustion of methane gas, which is released by rotting vegetation and which accumulates in swamps and marshes. Now it is known that fungi are responsible for those luminescent displays."

Now it was Charles' turn to snicker. "I'll bet you got a lot more of that luminescent fungi in Woomera after those atomic bomb tests. Talk about ridiation. Even the ground glows after dark."

"An obvious exaggeration but your point is well taken. However, the land around Woomera is far too arid to support bioluminescent fungi. If I were to hazard a guess – "

An aborigine woman appeared in the doorway. She stood silently with a quirky expression on her face, like the half smile on Mona Lisa. She wore a red paisley dress and leather sandals.

"Come in, Rosalie," said the professor. "Do come in."

"My lidy friend," Charles explained, as he rose to escort her to an empty chair. He gave her a peck on the cheek.

Rosalie looked uncertain. Instead of acknowledging Charles' kiss, she turned to Professor Wayward. "You must come, Doc Warren. You must see." Her accent was unlike that of Charles or the Professor – kind of a combination of British and Australian with a twang of Aborigine thrown in.

"What is it, My Dear?"

"Something very strange. You must see."

The professor did not hesitate. "By all means, Rosalie."

As far as I was concerned, strange meant bad. I

raised my eyebrows at Arlene. She put the last bite of her cake on the plate in front of her, and gulped. We all followed Rosalie out the door. She led us along a path that wound around the burnt remains of the schoolhouse. The ash inside the cinderblock foundation was cold.

Rosalie indicated a shallow depression that extended from the north wall of the schoolhouse to a nearby garden. Historically the Aborigines were nomadic, moving about the countryside in search of food. Nowadays, the Aborigines who lived in villages cultivated local crops to either eat or trade. Yams and bush onions were growing in the garden in front of us. Beyond the garden stood a grove of macadamia trees, and beyond that stood a grove of desert figs. A cornfield spread to the east. Corn was not indigenous to Australia, but British settlers had imported it long ago, and now it was a common staple.

The sun was sitting low on the horizon, and casting long shadows that outlined the western perimeter of the depression.

"See the stems? See how they are withered?" Rosalie stood on the edge of the depression, and pointed to the plants.

The flat leaves were brown; the slender stems were grooved and desiccated.

"How very odd," mused the professor.

The depression was only an inch or so deep. It would hardly have been noticeable had it not been for the border, which appeared as a slightly raised lip. Otherwise the ground looked normal. The plants seemed to be dying.

"Could the saltwater have drained this way from the schoolhouse, saturated the aquifer, and poisoned the plants because of its salinity?" I wondered.

"The ground is level and the bedrock is deep. If the saltwater did not sink straight down, I would think that it would have spread evenly away from all sides of the schoolhouse."

"What about chemicals in the laboratory?" offered

Arlene.

"Certainly there were toxic chemicals but not in large quantities."

"Something in the Mold?" Charles suggested. "Or if the Mold itself wasn't toxic, perhaps the residue became toxic after the Mold was ionized or burnt."

"That is a possibility." The professor stroked his beard. "The visible portion of these plants appears to be decomposing. As if the roots . . . "

Charles reached behind Rosalie and pulled a hoe out of a wheelbarrow. He reached into the depression with the implement, and snagged a withered plant. Instead of the yam coming out of the ground, the visible portion fell over as if it had been disconnected from its roots. Charles dragged the plant close to the edge. He twirled the hoe so the stems wrapped around the blade, then pulled in the handle until the plant was close enough to inspect.

The professor bent over and peered at the truncated stem.

Despite the fading light, from where I was standing I could see a tiny ball of white fluff on the nether end of the stem.

A collective gasp escaped from all of us.

"Mold!" I shouted.

"Right on."

"Oh, no."

"Oh dear."

Only Rosalie was mum.

The professor gazed over the garden. "The Mold has impregnated the soil."

Now I took notice of the distant groves. Instead of bright green, the leaves on the trees were dull and wrinkled. The ground was depressed around the slender boles.

"It must be growing through the dirt," Arlene declared.

"Of course," said the professor. "It must have extended some exploratory tendrils underground beneath the foundation. Topsoil consists largely of organic material: roots, seeds, grubs, burrowing insects, earthworms, microbes . . . "

I was staggered by the potential for global catastrophe. "So it's eating its way through the ground like a big smorgasbord."

The professor nodded. "Somehow the Mold sensed that there was more digestible matter in this direction than in any other . . . "

I felt queasy and suddenly helpless. "What do you mean, it 'sensed' it?"

"Not only that, but it stayed out of sight, as if . . . "

The implication was hideous. "Are you saying what I think you're saying? That the Mold is smart enough not to give itself away?"

"I don't know. . . . But it seems that way. However, appearances in nature are often deceiving. Protective mimicry, for example, is natural camouflage. Many animals imitate their background or other animals as a means to fool their predators: the walking stick resembles a twig; the wings of the viceroy butterfly exhibit a pattern and color design that is similar to that of the monarch butterfly, which is distasteful; the scarlet snake is a look-alike of the coral snake, which is venomous; many tropical fish have coloring that resemble the reef that they inhabit.

"There are a plethora of examples of happenstance in evolution. But there is no proof that any of these creatures *know* that they are camouflaged."

"Professor!" Arlene's shout brought the professor out of his scientific reverie. "Can we save the science debate for later, and deal with the situation at hand?"

My sentiments precisely.

"Right on."

Arlene pointed downward. "The depression is growing wider. The Mold must be expanding."

She took a step backward. The rest of us automatically followed suit. I saw that the rim of the depression was spreading outward by extremely slow degrees. Granules of dirt rolled down the tiny slope as the surface of the soil was undermined.

Charles took a spade from the wheelbarrow, and stabbed it into the ground next to the depression. He dug up a clump of dirt that was partly uninfected and partly infected. He slammed the flat of the spade against the ground so that the dirt spilled over the metal edges. Professor Wayward crouched and inspected the soil without touching it. After several seconds, he took a pen from his smock pocket and used it to part the dirt.

Ants danced through the uninfected part. As the pen carved a groove into the infected part, I saw a conglomerate of white strands engulfing one end of an earthworm.

The professor dropped the pen as if it had suddenly caught fire. "This is bad." He stood up slowly and surveyed the land in the distance. "This is very, very bad."

"This is bloody terrible," Charles expostulated. "The ground is impregnated with Mold, and it's spreading while we're working our jaws. What can we do?"

A macadamia tree toppled to the ground with a crash. In the distance, two fig trees took a decided list as their roots were consumed by the Mold. The two fig trees chanced to lean against each other, each using the other for support.

My mind was racing but I was speechless.

"We're all out of salt," Arlene shouted. "Can we burn it out?"

Charles tossed the spadeful of earth into the depressed area. "Dirt doesn't burn. We'd have to saturate the ground with petrol, and we're all out of that, too."

"But we've got to do something or the Mold will keep feeding and spreading."

"The underground biomass . . . " The professor shook his head. "We do not have the resources at hand to deal with the situation." He was losing his scientific detachment. "We need help."

"Help is a long way off," I said. I wondered what resources were available at a sleepy town like Alice

Springs to deal with an emergency that was beyond human experience. "Who's even going to believe that we can't eradicate a simple Mold?"

"This Mold is anything but simple," the professor ruminated. "Yet it is not invulnerable. We have already established two methods of liquidating it: saline solution and intense heat. Undoubtedly it has other weaknesses that we could exploit, given sufficient time."

"Now is not the time to experiment," Arlene said emphatically.

"Quite right, My Dear, quite right. We must get word to the military authorities. They are equipped to handle extraordinary matters. And we must do it at once. Every minute we delay enables the Mold to expand. Its growth rate appears to be exponential."

"The closest telephone is in Alice Springs," Charles noted.

"Don't the Aborigines have . . . smoke signals?"

"They're not American Indians, you idiot." Arlene spoke out of reflex. She instantly regretted her snide remark. "Tim, I'm sorry . . . "

I ignored her and her offhanded comment. I knew the kind of strain that she was operating under. "All right, then let's drive to Alice Springs."

"How?" Charles protested. "The Land Rover's petrol tank is empty. We used all the petrol to cremate Shane's body and burn down the hut."

"Siphon gasoline from the generator," Arlene suggested.

"Good idea!" I remarked.

Charles shook his head. "What good would that do? The bridge is out and the ford is flooded."

I nodded perfunctorily, but a crazy notion was forming in my head.

"This area is not safe. We might all become infected like poor Shane." The professor gazed at the orchards, where trees were now falling with reckless abandon. "We must evacuate at once."

Rosalie started. "I will alert the people." She trotted toward the village like a spooked doe.

"I've got an idea," I said hesitantly.

"It better be good, Mr. Biker. We don't have any options right now other than running away." As an afterthought, "Chased by a Mold."

I tried to put my thoughts together coherently.

"Let's back away some more," Arlene suggested.

The sound of falling trees was louder. We all took half a dozen steps backward.

"The bridge is out only to vehicular traffic," I explained. "I mean, the planks were washed away but not the beams – "

Charles interjected, "The Land Rover can't drive over the beams. Not that I wouldn't be willing to chance it," he was quick to add, "but the beams aren't spaced the same as the wheels."

"That's not what I meant. If you can get me to the bridge, I can walk across the beams. They're eight inches wide."

"And then what? It's twenty miles from there to Alice Springs."

"I can run that distance in under two hours. Non-stop."

Charles glared at me as if I had two heads. "You're daft. The heat has gone to your brain."

"Tim, this is no time to joke around."

"No, really. I ran cross-country in high school and college. Next week I'm supposed to compete in the New York Post Collegiate Marathon."

Arlene was stunned by my revelation. "I – believe – you."

Charles looked at her hard, then at me. "I'll get the hose."

"I'll change into my jogging outfit."

In the fading light of the setting sun I switched from my outback garb to shorts, T-shirt, and jogging socks and shoes. As I climbed out of my tent, the chill in the air raised goosebumps on my skin.

"Are you sure you can do this?" Arlene put her hands on my arms and rubbed away the goosebumps.

I was confident of my ability. "No problem."

Charles charged past us holding a can of gasoline, careful not to let any of the liquid slosh over the rim. Arlene and I followed him at a swift pace. Charles held a funnel over the Land Rover's filling tube, and poured gasoline into the wide metal opening.

Arlene insisted, "I'm going with you as far as the bridge."

"We might not make it that far. Depends on how much gas was in the generator tank."

"Then I'll walk back with Charles. Or wait at the bridge for reinforcements."

"There's no guarantee that I can get help in Alice Springs. I may have to hang around while the local authorities call the capital. And even then – no one may believe the wild tale I have to tell them."

"Colonel Ampers will believe you."

I didn't hear the professor's approach. "What makes you think so? This is a pretty bizarre story. And who is Colonel Ampers?"

"He is in charge of the army unit that is being established in Alice Springs now that Woomera is being abandoned as a testing facility. You will find the new base – such as it is – at the east end of town, before you reach the paved road. The dirt road leading north to the base is unmarked, but it can be distinguished by the deep tracks that numerous three-axle trucks have made in the sand. Tell him that you have an urgent message from me. The password is Code Red Zulu Red. He will understand. Explain the situation. He will muster all the forces that are at his command."

I ruminated for a moment. "We can't shoot the Mold to death. And artillery shells or bombs would only spread surviving strands across the countryside."

"Tim's right." She grappled with the professor. "Dirt won't burn, but if we saturate the soil with gasoline, we can ignite an underground fire that will incinerate the Mold from the bottom up."

"Like a peat fire," I remarked.

"Excellent suggestion, My Dear." To me, "A tank truck is kept at the base to refuel the vehicles."

Charles said, "Those heavy-duty army trucks run on diesel fuel, not petrol."

"Diesel fuel is ideal for our purpose. It is less volatile than petrol and therefore does not evaporate as rapidly. This will enable a greater quantity of the product to soak into the soil instead of vaporize into the atmosphere. Diesel fuel is denser than petrol, and releases more energy per gallon. Diesel fuel is also immiscible with water, whereas petrol becomes saturated and then cannot be ignited. Furthermore, diesel fuel is safer to handle because, unlike petrol, it cannot explode."

"All right, already," I shouted. "I get the point. I am forever beholden to Rudolf Diesel for inventing his internal combustion engine."

"Yes. Quite right. I apologize. However, diesel fuel cannot be ignited under normal atmospheric pressure because it has such a low octane rating, which equates to a high flash point. It must be compressed before it can be ignited in an engine."

"So what good is it to us?"

"Diesel fuel can be ignited if its temperature is high enough."

"How about mixing it with petrol?" Charles suggested. "The petrol fumes will burn. Will it burn hot enough to ignite diesel fuel."

"Yes, I believe it will. Very good, Charles."

Charles grinned.

"Okay, so that's settled," I said. "And the password is Code Red Zulu Red."

"That is correct."

Charles was enthusiastic but still uncertain. "But with the bridge out and the ford flooded, how is a tank truck supposed to get here?"

The professor took a deep breath. "I cannot say. But Colonel Ampers is like Hannibal when he crossed the Alps with elephants: he will either find a way, or make one."

Chapter 18

I was too lightly dressed for a twilight ride in the desert air. Arlene found a jacket in the back of the Land Rover; she tucked it around my shoulders.

"Charles, how will I ever find my way on all those forks? There were dozens of them, some with three tines."

"No problem going back, Mite. All the forks point awiy from Alice Springs. You'll be running toward the splits all the wiy. It'll be as easy as flowing down a funnel."

"All roads lead to Rome," I quipped.

"Right on."

Arlene wanted reassurance. "Are you sure you can do this?"

"I've done it before," I reminded her. "You think this is my first marathon?"

"Well, I – "

"My only worry is wild dingoes."

"No worry there, Mr. Biker. They only attack small animals."

I put on a sense of bravado. "If I see any dingoes with fire in their eyes, I'll spit in their eyes and put the fire out."

Arlene managed a weak smile.

Charles brought the Land Rover to a halt at the end of the bridge over Noname Creek. The din of rushing water was deafening. Even in the failing light I could see the raging torrent that splashed over the support beams. The wet wood would be as slippery as the dickens. We piled out of the vehicle.

Charles thrust a canteen into my hand. "Quaff this down."

I tanked up with water in preparation for the run of my life – and possibly the life of everyone else on the planet. "I wish I had some chocolate for instant energy."

"I have some in my purse."

"You brought your purse?"

Arlene looked stunned. "Of course. A woman never travels without her purse."

Charles took back the empty canteen. "You have chocolate?"

"Of course. A woman never travels without chocolate."

I rolled my eyes, but I took the proffered chocolate. After I gulped it down, Charles handed me another canteen. I swilled some water in my mouth, then chugged as much water as I could stomach. I handed the coat to Arlene.

"As they say in the insane asylum, I must be off."

Arlene leaned forward and gave me a peck on the cheek. "Be careful, Tim."

Charles slapped me on the shoulder. "Good luck, Mite."

I nodded silently. I took a deep breath and placed my right shoe on a beam. I turned on the ball of my foot to test the traction. The beam was wet but not – god forbid – slick with mold. I slid one foot in front of the other like a tightrope walker until I gained some confidence in the friction that the planed wood provided. Water splashed lightly over my shoes and wet my bare legs. Then I walked normally until I reached the pier in the middle of the creek.

Waves crashing against the upright structure sprayed water higher than my head. A rogue comber struck my upraised foot and spun me partly sideways. I twisted around with one foot in the air, balancing on my other foot. I hung that way precariously for what seemed like minutes, but which couldn't have been more than a second or two. I heard Arlene scream. Then I regained my equilibrium and reacquired solid footing. From that point onward I reverted to the tightrope walking technique.

I reached the other side of the bridge without much difficulty. After all, the beams were eight inches wide: more like a sidewalk than a tightrope. I shouted and waved from solid ground. In the fading light I could see

Arlene and Charles waving back at me, but if they were calling I could neither see their mouths move nor hear their voices. The noise of the floodwaters drowned out all other sounds.

I did a few simple leg-stretching exercises: not as many as I would do ordinarily because of the time bind, but I knew that if I did none at all I would be hurting long before I crossed the finish line.

I decided that heroic efforts were not in order. I couldn't afford to collapse at the finish line or be crippled after the race. My competitor was not another runner but the expansion of the Mold. I started at an easy lope for the first mile or so. Then I increased my pace until I got my second wind. At that point I slowed down a tad, to a speed that I could maintain for the distance without unduly damaging myself.

Long-distance running is partly a matter of pacing and partly a matter of psychology. If you start by running as fast as you can in order to keep up with the competition, you run out of steam before the end of the race. If you run too slowly, you finish in fine form but don't win the race. The trick was to know your best pace, maintain that pace, and ignore your jeering competitors. Many times I have been outrun in the first half of a race, only to pass exhausted runners in the second half.

A smart runner knows that he is racing against himself, not against other runners.

Cross-country running differs only in that one must take the terrain into account. Cross-country courses include hills to break the monotony. When ascending or descending hills, it was crucially important to maintain your breathing rate, not your speed. If you tried to maintain your speed during an ascent, you found yourself breathing too hard and wearing yourself out. In this case, the trick was to slow down so that your breathing rate remained the same.

Once I hit my stride, I stuck to it. The road to Alice Springs was essentially flat. The few fluctuations never rose or fell more than a few feet on a barely perceptible slope. My biggest obstacles were patches of soft loam that punctuated the hard-packed sand. The recent rain made the roadway mushy and clingy. When my shoes sank into soft spots, I felt as if someone had installed brakes in my soles. I soon learned that when I encountered loose or wet sand, it was easier to jog along the untrammeled land adjacent to the road, where tires had not churned the earth, and where the spreading roots of sparse vegetation kept the ground firm.

The light blue of twilight darkened gradually to deep purple, then finally to pitch black. I soon lost all sense of direction. My only guide was the pair of tire ruts that carved a path through the bleak desert. I kept glancing over my shoulder for the Moon above the eastern horizon, but it had not yet risen. Stars shone brightly between patches of dark clouds that portended more foul weather. Once I spotted the Southern Cross among unfamiliar constellations, but I didn't know its bearing with respect to the compass.

Charles was mostly right about the forks in the roadway. Generally they were shaped like a capital Y. Occasionally, however, a fork looked more like a capital T than a Y. Then I had to stop and study tire tracks in the starlight in order to determine which extension was the upright of the T.

I stumbled frequently on soft spots in the sand, and danced sideways until I regained my equipoise. Once I jumped over a small furry animal that darted across the road in front of me. It gave me quite a start until I remembered reading in an airplane brochure about the potoroo, or rat kangaroo, so named because its miniature stature reminded early settlers of the rodent after which it was named.

The night air was as crisp as refrigerated lettuce. Despite the damp chill, my T-shirt soon was soaked with sweat, and beads of perspiration dripped down my forehead into my eyes. In order to conserve energy, I jogged slightly slower than my racing pace. My breathing rate didn't change as I followed the bends in the road for more than an hour.

Ahead, the utter darkness was dispelled a wee bit by a pinpoint of light that adorned the horizon somewhat to my left: about forty-five degrees to my direction of travel. As the light grew steadily brighter, I began to think that a vehicle was striving to cut me off at the pass. I stopped for a moment to align my sights. The light didn't advance, retreat, or waver. Fifteen minutes later, as the road angled progressively to the left, I came to the abrupt realization that I was seeing the streetlights of Alice Springs.

Now I ran straight for the town on a road that veered moderately to the left and right in order to avoid the worst geological hazards: hillocks, gullies, washouts, stands of shrubbery, and clumps of grass. Alice Springs lay dead ahead.

I maintained my stride but reduced my pace until I realized that distances were deceiving in the open desert. The town must still have been more than five miles away instead of right around the proverbial corner. I picked up my pace again until the single bright light resolved itself into multiple points of illumination. By that time I was nearing the end of my resolve. My chest muscles compressed my lungs effortlessly for each exhale, but my calf muscles and upper thighs were aching from the uncommon effort of jogging through soft undulating sand instead of along a flat and level street in the City.

I couldn't breathe a deep sigh of relief at the sight because I was already breathing to near full capacity. With the finish line clearly in sight, I instinctively increased my pace and swung my arms all the harder, adding a last-ditch burst of speed.

Suddenly I stumbled onto a hardtop road surface. I didn't stop, but I slowed to a walk and turned a few circles as I strived to recall my instructions. I had passed the cutoff to the army base. Having thought that I was nearing the end of my goal, I found myself drained of energy and lacking the strength to retrace my steps at my previous pace.

I walked in circles for a minute or so, then ran back

the way I had just come. I kept my eyes peeled in the darkness for lights to my left on the outskirts of town. With the MacDonnell Range as a backdrop, the ground was a black as the inside of a pocket. Because the sky was now largely overcast, only a few stars showed in the firmament. Thus I felt myself extremely lucky to notice two deep ruts that veered northward into the desert.

I jogged along the ruts for half a mile to the top of a low rise. As soon as I crested the hill I spotted a handful of lights perhaps a mile or so away. By this time I was all but spent. It must have taken me six or seven minutes to run that last mile. The first building I passed was a wooden shed that could have been a guard shack or an outhouse; it showed no lights or signs of habitation. I stopped running to tug on the door. It opened easily.

"Anybody – home?" I gasped.

There was no answer, so I walked toward a cluster of wooden buildings of which several showed an incandescent glow in their windows. My idea of an army base was a stockade of wooden posts or a crosshatched fence that acted as the first line of defense, punctured by a fortified gate that was guarded by military police who wielded rifles and side arms.

I was still breathing hard when an unarmed but uniformed soldier stepped out of an unlighted building which, judging by the odor, definitely *was* an outhouse.

He must have heard my rasping breaths and shuffling footsteps, for he halted sharply as if he had run into a brick wall.

"Didn't mean – to startle – you," I gasped. "I've got – an urgent message – for Colonel Ampers."

"You give me a start, Mate. We don't get many visitors out here, even in the daytime." He didn't appear to be alarmed or suspicious. "Did you run here from town?"

"Farther." I was getting my breathing under control. "I've got to see the colonel. It's important."

"Sure thing, Mate." He flung his arm to the right

and indicated a nearby building with one finger. "His light's on, so he'll still be working. Go right on over."

"Thanks." I tossed off a halfhearted salute and limped to where he was pointing. I passed a shed from which the steady thrum of a generator emanated. The next building was a single-story cabin. A pale yellow light escaped through a screened window and screen door. I knocked on the door.

"It's unlocked."

I was still breathing heavily when I stepped inside. The room was sparsely furnished with a wooden desk, a couple of wooden chairs, and a row of metal filing cabinets. A lone shaded lamp illuminated a sheaf of papers on the desktop. The straight-backed man who sat behind the desk reminded me of a British commando from World War Two. His face was firm and his body was lithe. I imagined that he was as strong as a bull and as hard as a ten-penny nail. Brown bushy eyebrows matched the color of his crewcut hair.

"Colonel Ampers?"

"Yes." He looked up and down at my unusual state of dress. "What do you want?"

By now I was able to string my words into a sentence without gasping for air in between syllables. "I have a message from Professor Wayward."

The colonel stood and strode around the desk. His eyes glinted, and a glimmer of a smile touched his lips. "How is the old fellow?"

"Sir, the message is urgent. He told me to give you the password Code Red Zulu Red."

The colonel's expression sagged instantly and he halted in his tracks. He stared at me as if I had just struck him in the forehead with a brick, or called his mother a nasty name.

"Would you – would you repeat that please?"

"Code Red Zulu Red," I pronounced slowly and deliberately. "The professor said you would know what it meant."

Without another word, the colonel spun around and leaned over his desk. He grabbed the phone, turned the black plastic base toward his flat belly, yanked off the handset and snapped it to the left side of his face, then dialed a number with a rigid index finger. Judging by the number of clicks that it took for the rotary to return to zero, the phone number contained several eights and nines.

"This is Colonel Ampers at Alice Springs. I have just received a Code Red Zulu Red. Mobilize immejately. I will call back with further instructions." He placed the handset in the cradle, held it there for two or three seconds, then picked it up and dialed another number. "Sergeant Gompton, we have a Code Red Zulu Red. Get over here immejately."

When the colonel turned to face me, I saw that his chest was heaving an infinitesimal amount quicker. Now his eyes lingered over my abbreviated attire. "You've come a long way, Mister."

"Yes."

"What is your name?"

"Baker. Timothy J. Baker."

"American." He stated it flatly, as an obvious fact. "Where is Professor Waiyward?"

The colonel pronounced the "a" like a diphthong: halfway between a long "a" and a long "i."

"At his schoolhouse."

He affected a single nod. He turned and ducked through a doorway into the back room. He reappeared a moment later wearing a military bush hat and bearing a bundle of olive drab clothing in his arms. He handed me an olive drab towel. "Dry yourself off, Mr. Baiker."

"Thanks." I did as he suggested, and ran the towel over my exposed skin.

"Take off that wet shirt or you'll freeze to death."

I did so. After I dried my chest and back, the colonel handed me an olive drab T-shirt. I was grateful for the way the cotton felt against my skin. My pores were still producing sweat, but not so much that the T-shirt couldn't soak it up.

"I'm a little taller than you but these will have to

do." He handed me a pair of fatigues like the ones that he was wearing. "My boots are too big for you."

I started to don the fatigues. "My jogging shoes are okay."

After I shrugged into the shirt and pulled the trousers over my jogging shorts. I rolled up the shirt-sleeves and trouser cuffs. He handed me a belt with a brass buckle that shone like the noonday sun. I was fussing with the buckle slider when the screen door burst open.

"Sir?"

"Sergeant, this is Mr. Baiker. He brought the message from Professor Wayward."

The sergeant nodded. "Yessir."

"Mr. Baiker, tell me what the professor needs. You can fill me in on the nature of the emergency later."

"Well, we thought – that is, the professor thought – that we could incinerate the – " I hesitated at saying Mold because it sound so improbable. " – the infected area. He thought you might have a tank truck loaded with diesel fuel."

Colonel Ampers glanced at the sergeant.

The sergeant acknowledged without formality, "About three-quarters full."

"Do you have any gasoline – er, petrol?"

"Jerry cans for the generator. Fifteen or twenty."

"The professor said that you have to mix gasoline, er, petrol, with diesel fuel to get it to ignite."

"That's correct."

"Then you'll have to dump the spare gasoline, er, petrol, into the tank truck to make a burnable mix."

"If you say so."

I was thinking furiously. "How about flamethrowers? Or napalm?"

"This is an advance administration base, Mr. Baiker, not a combat arsenal."

Now it was my turn to nod. "Do you have anything else that will burn?"

The colonel and the sergeant raised eyebrows at each other.

The sergeant: "There's propane for the space heaters."

The colonel: "And kerosene for the lanterns."

"Whatever you've got."

"Sergeant, see to it."

"Yessir." The sergeant nodded and left.

"What else, Mr. Baiker?"

"Well, do you have any way of spreading the inflammables? Portable water pumps? Insect repellent canisters? Anything. We need a large-scale spray. Over several acres of land."

"Would a fire engine suffice?"

"Perfect."

"There's a hose truck in town." Colonel Ampers scooped up the phone and dialed zero. "Get Chief Constable Harris. . . . Find him and have him call Colonel Ampers immejately." He hung up the handset. "What else?"

Events were moving so fast that my head was spinning. The colonel was delivering combustibles faster than a clerk at Gino Marchetti's could serve fifteen-cent burgers and French fries.

"Well, like I said, anything that will burn. We need to saturate the ground and incinerate everything down to the roots. To bedrock if possible."

The colonel nodded once. His eyes never wavered from mine. I couldn't think straight when I was staring someone in the eyes; I had to shift my gaze to the upper left or right. But his mind raced like a steam engine on full speed ahead.

"There's a petrol staition in town. We can fill the empty jerry cans . . . " He was still mulling over possibilities when the telephone rang. "Yes . . . We have an emergency situaition. . . . We need every man you can get: constables, firefighters, and anyone you can deputize or volunteer. . . . And we need the hose truck. . . . I want you to dump the water and fill the tank with petrol. . . . Yes, I said petrol. . . . You bloody well better do it. . . . If we ruin the truck, the army will pay for a new one. . . . Immejately. . . . Meet me on the east road."

I swallowed hard. I had no idea what Code Red Zulu Red meant, but it obviously was a protocol of a most serious nature. In the span of a few minutes, the mere pronunciation of the phrase had instigated a reaction of massive proportions. And the colonel didn't even know what it was all about!

"Excuse me, colonel, but – don't you need some kind of authority to, well, to declare war?"

"Code Red Zulu Red *is* my authority. You bloody Americans are too wrapped up in bloody red taipe to ever get anything done. Here in Australia, we don't talk about doing. We *do*! Now follow me."

He rushed through the screen door like a bull charging a red flag. I followed at his heels. I practically had to run to keep up with him. We met the sergeant and a gathering of troops. Flashlights indicated the flurry of activity which the professor's four short words had fomented. Diesel engines rumbled and gasoline engines purred. The dirt roadway was crowded with vehicles: a tank truck, two canvas-topped troop carriers, and two Land Rovers.

Soldiers ran back and forth in what looked like pandemonium but which was actually a rehearsed and carefully coordinated battle plan. The sergeant shouted orders left and right.

"You stick with me, Mr. Baiker. You're the leader of this parade."

"Me?"

"Yes, you." The colonel speared me with a pair of baleful eyes that shone like first magnitude stars in the surrounding darkness. "I don't know where the professor's bloody schoolhouse is located. You have to lead us there."

"But – I don't know how to get there."

"Didn't you just come from there?"

"Well, yes. But I don't know how to get back. The road is full of forks."

It sounded stupid when I said it like that, as if I were describing a cutlery set, but the colonel didn't notice.

"Didn't you just drive here on those forks?"

"Well, no. Not exactly. I ran here from the bridge."

"You ran? From Damifino Canyon?" He raised his eyebrows as if he suddenly understood why I arrived dressed the way I was. "I ought to put you in charge of physical training of my troops."

"Charles the safari leader – " I didn't remember how to pronounce his Aboriginal surname. " – drove me out there."

"So where is he?"

"On the other side of the bridge."

"Why in bloody hell didn't he drive you all the way?"

"Did I forget to mention that the bridge is out?"

"You bloody well did *not* mention that fact. What else do I need to know?"

"Well, the ford in Noplace Valley is flooded."

"Jesus bloody Christ." For the first time in the few minutes since I had met him, the colonel was speechless. He whipped his hat off his head and ran bent fingers through his close-cropped hair. "What do you mean by 'out'?"

"Well – "

"Stop saying 'well.' Just say what you bloody well mean."

"Well – that is, uh – The pier and abutments are still in place. And the support beams, too. The basic structure is fairly sound. But the floodwaters punched up the road planks and washed them away."

He mulled this over for half a second. "Sergeant!"

"Yessir."

"Gather all the spare boards for the new barracks and load them into the trucks. And every keg of nails. We've got a bridge to rebuild."

"Yessir." The sergeant barked out orders, and men jumped to his command.

"Anything else, Mr. Baiker?"

"I think that's about it." I tried to avoid his awful stare so I could think. "Wait a minute. The Enders know the way to the schoolhouse. And they're in town. Or they should be."

"Sergeant!"

"Yessir." I don't know how the sergeant always managed to reach the colonel's side so quickly. He was never more than two seconds away.

"Call Constable Harris and have him track down the Enders brothers. They're to meet us on the east road."

"Yessir."

Colonel Ampers personally inspected the tank truck and troop carriers. He stepped aside as soldiers tossed boards onto the bed between the bench seats. He performed a radio check on the radios in each of the Land Rovers; both were working.

I heard a siren wailing in the distance. The fire engine was on its way. I glanced at my watch, and by the position of the phosphorescent dials I noted that barely half an hour had passed since my precipitous arrival. This was mobilization with a capital M. What chance did the Mold have against a colonel who was ready to storm Iwo Jima without hesitation and at a moment's notice?

"No rifles, Sergeant. Hand guns only." The colonel himself was unarmed.

"Yessir."

Colonel Ampers jumped into the driver's seat of the leading Land Rover. "You're riding shotgun, Mr. Baiker, as you Americans call it."

"Okay." I climbed into the passenger seat like a cowboy climbing onto a stagecoach in a Western movie.

The sergeant sat on the bench seat behind us. "All ready, sir."

"Call Woomera and tell them we are on the way. I will give them a sitrep as soon as I have anything to report."

The sergeant called on the radio that was strapped to the rear seat. His report was clear and succinct, although most of the acknowledgement was undecipherable because of static. Behind us, the building lights dimmed, then were extinguished completely when the generator was shut down. No one was left in the army base.

All four tires spouted rooster tails as the colonel raced along the dirt road in four-wheel drive. In less than a minute we spun onto the main road to join the rest of the convoy: a red fire engine with "Alice Springs" emblazoned on the sides in white paint, the Enders' pickup truck, and a dozen assorted vehicles whose engines idled loudly. I pictured an Alice Springs that was abandoned to the roos. Had I left the town in a different reincarnation or only a few days ago?

A man appeared next to Colonel Ampers. "How do you want to proceed, Colonel?"

"Thanks for your promptness, Mr. Harris. Have the Enders brothers lead the way. And tell them not to dawdle. My vehicles will go next."

"Understood." The man disappeared in the darkness.

A minute later, the Enders' pickup cruised past us. Colonel Ampers let the truck get about a hundred yards ahead before starting after it. That was enough distance for the dust to settle. It must have been standard operating procedure in the outback, for no vehicle strayed closer than that to the vehicle in front of it.

I slumped in the seat, glad to relax my sore leg muscles now that the responsibility of effort was taken out of my hands and put into the capable hands of Colonel Ampers. A wave of relief washed over me. With the amount of "fire" power behind us, I didn't see how we could fail to frizzle the Mold to extinction. We could easily burn down all of Rome while Nero played on his fiddle.

This was the beginning of the end. Or was it?

Chapter 19

During the drive I told the colonel and the sergeant practically everything that had happened since my arrival. They accepted every word as gospel. I kept mum about my professional rivalry and interpersonal conflict with Arlene. They didn't need to know about that.

"Ordinary salt water kills it," remarked the sergeant, who had introduced himself as Gompton. "How odd."

"*I* don't understand it. The professor said something about the valence of sodium and chlorine atoms, and the strong ionic potential that was produced when common table salt was dissolved in liquid. He speculated that other mineral compounds might achieve the same effect: other chlorides as well as sodium compounds such as sodium nitrate."

"We've got lots of that at Woomera," the colonel admitted. "It's used in solid rocket propellants, and in the manufacture of conventional explosives. Sergeant, tell Woomera to send all the nitrate they can muster."

"Yessir." Sergeant Gompton called on the radio, but all he got was static. He extended the retractile antenna as far as it would go, and stuck the end out the rear window. "Nothing. Too much interference."

"Try again in five minutes." The colonel explained for my benefit, "Sunspot activity. It has reached its peak and is supposed to last for several days. It plays hell with radio waves."

I comprehended his meaning. A sunspot was an area of reduced temperature but intense magnetic activity in the sun's photosphere, or outer layer. The cause or origin of sunspots was unknown.

During times of intense sunspot activity, the Sun emitted large quantities of alpha particles and beta particles at nearly the speed of light. What an astronomer called an alpha particle, a chemist or nuclear physicist called a helium nucleus: two protons and two neutrons in contact with each other. A helium *atom* consisted of a nucleus that was circled by two electrons whose negative charge balanced the positive charge of the protons; neutrons had no charge – they were neutral.

On the atomic scale, alpha particles were huge in comparison to beta particles: more than seven thousand times the mass. A beta particle was a high-speed free electron, perhaps one that had been stripped from an atom by the intense heat or radiation found within grand old Sol.

When these charged particles struck the Earth's ionosphere, their interaction with high-flying atoms and molecules released energy in the longer wavelengths of the electromagnetic spectrum. The Earth's magnetic field channeled these interacting particles to the poles, where the interactions were concentrated – like water pressed together at the bottom of a funnel. Some of the energy was released in the form of visible light, which created a phenomenon that was known as – voilà! – an aurora.

In the northern hemisphere the auroral display was known as the northern lights, or aurora borealis. In the southern hemisphere the display was known as, appropriately, the southern lights, or aurora australis. Put that in your pipe and smoke it, Professor!

I instinctively looked southward. The clouds had temporarily dispersed close to the horizon, but we were too far north to see the wavy green curtains that an aurora manifested. Back home, there was too much light pollution to see the aurora borealis from the Big Apple (as jockeys and jazz musicians used to call New York during the Great Depression), but I saw the display once from Halifax, Nova Scotia.

"Now let me ask you – why don't we bomb the bloody Mold out of existence?"

"Well – " As soon as I said "well" I realized my mistake. The colonel let me slide. "That is, our greatest fear is the Mold's amazing resilience to total destruction. It seems as if every strand has the ability to survive on its own. A bomb might kill the majority of the organism, but any tendrils that were not incinerated by the blast would be wafted by the wind to the far corners of the continent, where they would take root in the understory. Bombing the Mold would spread it rather than kill it."

"A handy defense mechanism," Sergeant Gompton commented dryly. "Or survival tool."

"Exactly. If the Mold isn't contained while it's small, it could take over the continent. It can't spread any farther because of the ocean. And it will die off once it has consumed every bit of organic matter in Australia. But most people would view that as an extreme means of extermination."

"Quite." The colonel accepted my satirical prognostication stoically.

"Humph," the sergeant grunted.

My witticism was out of place. This was not a laughing matter. Millions of lives were at stake.

"We have another extreme means of exterminaition that we would employ before allowing Austrailia to be overrun. Sergeant, call Woomera and tell them to prepaire for the worst."

"Yessir." Sergeant Gompton worked the radio. He twisted dials, adjusted the gain, and changed frequencies – all to no avail. "Nothing, sir."

"Try again laiter." To me, "How much do you know about Woomera, Mr. Baiker?"

"Only what I've read in the tourist brochures. It's a rocket testing facility, isn't it?"

"It used to be. Woomera was officially abandoned several months ago. Only the clean-up crew occupies the village. We are now in the process of moving the tracking station to Alice Springs."

"Tracking station?"

"Let me give you a bit of history that might have some bearing on current events. During the war, Nazi Germany blitzed London with V2 rockets that did some damage, killed a few people, and caused general havoc among the civilian population. The rockets served no military purpose. They did not attack targets of strategic value. The Nazi's objective was to demoralize England and thereby force capitulation."

Sergeant Gompton grunted. "That will never happen."

"Quite right. The Nazis didn't reckon on British determination. Instead of scaring off the Brits, the blitzkrieg served instead of firm their resolve. As Churchill said, 'Never give in. Never, never, never, never give in.' He spoke for every living Brit. Hitler was used to dealing with the French, who ran and hid at the first hint of attack. He failed to understand that the British were a different kind of people.

"After the war, everyone recognized the vast potential of rockets as ordnance delivery systems. If a primitive rocket could cross the English Channel, a more powerful and sophisticated model could cross oceans."

"The Intercontinental Ballistic Missile," I interjected.

"Quite. There was a mad scramble to establish rocket research bases and development facilities. The U.S. and the U.S.S.R. had vast tracts of empty land in which to destroy their mistaikes. England was densely populated. The Woomera Prohibited Area was established as a cooperative effort between England and the land down under, with a few Americans thrown in to maike life interesting. The WPA is as big as all of England, maiking it an ideal testing site with a long down-rainge corridor.

"Woomera rocket scientists concentraited on designing engines, throttle mechanisms, guidance controls, fuel feed components, and highly explosive fuels. Live testing commenced with the launching of small sounding rockets: sub-orbital rockets that carried a variety of instrument packages into the fringe of outer space, and loaded with telemetry equipment. Eventually, Woomera evolved into an aerospace tracking staition. It tracked Russian spaicecraft and satellites as well as American. Woomera was an essential link in the continuous tracking of John Glenn's orbital flight in

1962.

"You may have seen it on the telly," said Sergeant Gompton, "but we were in the control shack during all three orbits."

"Quite. Now Woomera has been shut down because of nuclear contamination."

"How did that happen?" I doubted that the colonel could see my beetled brows in the dark. "Were they experimenting with atomic fuels? Or uranium derivatives?"

"Not exactly."

The Enders' pickup veered onto another side road that was invisible to us. Colonel Ampers slowed down, spotted the narrow track as we came abreast of it, then spun the steering wheel and turned onto the narrow dirt pathway.

"It is difficult to hide an atomic bomb blast from the public. The mushroom cloud can be seen for scores of miles, and ground tremors are detected by seismographs all over the world. Once they get wind of things, inquisitive reporters stick their bloody noses into places where they don't belong. Nothing personal, you understand."

"I've had my nosed bloodied on more than one occasion. Figuratively speaking, that is."

"Quite. In a maijor press release, the government let it be known that Woomera was also being used as a nuclear weapons testing facility. England had already detonated a nuclear device on one of the Montebello Islands, off the coast of Western Austrailia, in 1952. This made the 1953 atomic blast at Woomera seem like a continuaition of that test."

" 'Seem like'?"

The colonel took one hand off the wheel and held it palm outward toward me as a gesture of restraint. "The blast was successful but left some residue, and I am not referring to nuclear fallout, although there was some of that too. A second blast a couple of weeks laiter cured the problem.

"A second incident occurred at a site in Maralinga

in 1956, again requiring nuclear remediation. Three additional bombs were detonated as a cover story. One of those bombs was dropped from an aircraft. More – "

"*As a cover story?* Are you telling me that the government detonated three nuclear devices as a *cover story?*"

"Quite. More devices were detonated periodically as bona fide tests, with great fanfare and open-air publicity, so the public could be advised that nuclear testing was all that was going on."

This story was getting weirder by the minute.

"And what *was* going on?"

"That is the point that relates to current events."

I couldn't fathom where this background material was leading.

"Let me inform you, Mr. Baiker, that Austrailia has a strict Official Secrets Act. You must not repeat any of what I tell you."

"Then why are you telling me?"

The colonel laughed. "I'm a good judge of character, Mr. Baiker. I trust you. But more than that, Professor Wayward trusts you. He dispatched you on a mission of the utmost importance to the safety of the world. He entrusted you with a top-secret activation code that can unleash tremendous destructive powers. You are deeply involved in this deadly situaition, and were it not for your diligence and active participaition, the world might possibly cease to exist as we know it."

Where had I heard that lethal phrase before?

"I leave you on your own recognizance, not as a reporter but as a person." After a pause, "And as I said, the Official Secrets Act is very strict. The government treats any violaition with extreme severity."

There was no detectable inflection or innuendo in the colonel's voice. I looked him squarely in the eyes, and he looked back at me just as squarely. A rustle in the back seat diverted my attention. I glanced at Sergeant Gompton. He didn't say a word but made a single nod in my direction. I noticed that his right hand rested lightly on the side arm in its holster. I looked again at the colonel.

"No red tape?"

"No red tape."

The outback was a big place. A person could go walkabout and never be seen again.

"I understand. And I appreciate the confidence you have in me."

The quiet time that passed seemed interminably long.

"Sounding rockets returned to Earth and were collected downrange. One that was carrying live organic substances suffered a parachute failure. It crashed hard, and debris was strewn over a large area, along with the scientific packages. The organic substances – microbes, bacteria, single-cell plants and animals – were dispersed by the impact that shattered their containers. The purpose of the experiment was to determine how simple organic structures were affected by cosmic raidiaition."

The eleven-year sunspot cycle immediately came to mind. I had done a piece on an obscure Dutch biomedical research team that found correlations between sunspot activity and influenza epidemics. They concluded that solar radiation caused mutations in flu viruses, creating new strains for which mankind had no natural immunity.

"Professor Wayward can furnish the biological details because he was in charge of the clean-up operations. He conjectured that the structure of one of the organisms was altered, or mutated. It infected the local flora which then grew out of control. The disease was too widespread to be destroyed by conventional means, although we certainly tried. Once we lost containment, the professor recommended gross extermination.

"After the first blast, soldiers and scientists crawled over the area with a fine-toothed comb. Some of them died shortly thereafter from raidiation poisoning. Many of those who accumulated lesser doses died later from various forms of cancer. Of greater immediate importance, some infectious particles that were retrieved from outside the blast site remained active.

"Although the second blast was reported in a press release to have been smaller than the first, in actual fact it had twice the yield. The second blast was powerful enough to eradicate the menace. The 1956 incident was a virtual repeat of the 1952 incident.

"Nuclear tests were discontinued last year, but the arsenal is being maintained in the event of another catastrophe resulting from near-space contamination. So you see, Mr. Baiker, if we don't nip this Mold in the bud – "

Or in the spore, I thought, *as the case may be.*

" – then we are prepared to exterminate it with extreme severity."

Chapter 20

The mind boggled. At least, mine did.

On one hand I found it difficult to accept that the Australians had used atomic bombs to destroy infected vegetation. It was like burning down a skyscraper to get rid of a few termites. On the other hand, if those previous pathogens were as infectious as the Mold – and I must believe that they were – then the measure didn't seem so extreme.

I was still pondering the lethal potentiality of Molds and microorganisms when we reached the bridge across Noname Creek.

I was half hoping that Arlene and Charles would be waiting impatiently on the other side of the bridge for my return. They were not.

Colonel Ampers and Sergeant Gompton wasted no time in organizing the enlisted men for bridge construction. Both troop carriers were backed up to the western abutment. Deputies and volunteers unloaded boards and nails, and stacked them neatly on the ground. Wood that had been planned originally to build barracks at the fledgling army base were laid across the bridge beams and nailed in place.

Local carpenters supervised the work and placed the boards where they belonged, but the soldiers banged most of the nails. No care was taken to keep the ends of the planks even with outer beams. Because the wallboards were longer than the width of the bridge, they stuck out raggedly from either side. Expedience was the watchword of the day – or night – not cosmetics.

I stayed out of the way, along with Edward, Albert, and Constable Harris.

In conversation I learned that Harris was not only the Chief Constable of Alice Springs, but also the head of the fire department and the driver of the hose truck. When he wasn't working in either of those capacities –

due to the paucity of crime and conflagrations – he spent his time as the wildlife resource officer: he collected and housed stray cats and dogs, tended injured wild animals, and ran the local pet store. He had more hats than a haberdashery.

Floodwaters were still raging. Everyone who ventured onto the bridge was soon soaked with spray. No one complained.

Clouds that were thickening in the nighttime sky threatened to burst their seams at any moment and drop another deluge. I wondered how well gasoline and diesel fuel would burn in a downpour.

"Feast or famine," opined Constable Harris in a soft, even voice. "It don't rain in the desert for months on end, then we get a gully washer all at once."

After the carpenters and troops laid planks all the way across the bridge, they started a second tier. Planks that were intended for the siding of buildings would not support the weight of heavy-duty trucks. The second tier was laid edge to edge like the first, but was offset so that the edges fell on the middle of each plank in the first tier. This was the same method that was used in laying bricks.

The nails weren't long enough to go through two planks and get a grip on the beams. As long as the first tier was firmly secured, the carpenters agreed that the second tier would remain fastened to the first tier.

The carpenters were uncertain about two tiers being strong enough to support the fully-laden tank truck. There weren't enough planks for a full third tier, so they staggered the leftover planks to create a partial third tier by leaving a space between each succeeding plank. Because they were running out of nails, the third tier was only partially secured to the second tier by widely spaced nails.

"Get that tank truck in position." Colonel Ampers never shouted. He didn't have to. He had a deep commanding voice that projected through the air like the twang of a base fiddle.

Neither did Sergeant Gompton need to raise his

voice. "Yessir."

"The hose truck will cross next."

Loaded as it was with gasoline, the hose truck was at least as heavy as the tank truck.

Constable Harris approached the colonel with concern. "Pardon me for making a suggestion, but what if the bridge ain't strong enough to hold these trucks. Shouldn't you test the crossing with the lighter trucks first?"

"That's a good idea, Mr. Harris. But we can't achieve our objective with Land Rovers and troop carriers. We need more than ill-equipped men to do the job we need to do. If we can't get the combustibles across, we might as well pack our kit bags and go home."

Constable Harris nodded. "Yes, I can see your point."

"If my man makes it across with the tank truck, this jury-rigged bridge should hold your fire engine."

The constable nodded again. "Yes, I suppose you're right." He stepped aside so the first truck could roll.

I could see that he didn't relish having to drive the fire engine across a bridge that had been designed for light vehicular traffic. Most of the discussion among the carpenters was not whether the triple-tiered planks would crack or splinter under the weight, but whether the central pier would hold its own or collapse.

It had already been decided that the driver's-side wheels should roll over the outer beam. The outer beams were thicker and stronger than the cross beams. By keeping half the truck on one of the main structural supports, the weight that was placed on unsupported planks would be reduced.

Half a dozen soldiers crossed the bridge on foot. One of them – a corporal – held a flashlight on the outer beam for the truck driver to home in on. After some backing and maneuvering, and after much discussion, the tank truck was aligned to everyone's satisfaction. The diesel engine rumbled.

"Pardon me for making another suggestion . . . "

Colonel Ampers eyed the constable smoothly. "Go

ahead, Mr. Harris."

"When we take four-wheelers out into the desert, we find that we get better traction in loose sand if we let some air out of the tires. The bottom of the tire spreads out and increases the surface area so the tires float better. I was thinking that we might spread the load the same way."

"Good idea, Mr. Harris. We'll do it. Serg – "

"Yessir. I'll get the men right on it, sir."

Sergeant Gompton issued orders with a few clipped words. After removing the caps, the soldiers used sharp rocks and knifepoints to depress the stems of the Schrader valves. The collective hissing was audible above the noise of rushing water in the nearby creek.

"Mr. Harris, would you check the tires for the proper deflation?"

"Will do, Colonel."

Harris inspected the bulge of each and every tire. He indicated whenever he wanted more air released.

"Right smart cookie, that," said Albert.

"Sharp as a razor, quiet as a clam," said Edward.

"I guess that makes him a razor clam." My weak attempt at juvenile humor went over the Enders like the proverbial lead balloon. After seeing their reaction – or lack of it – I was glad that Colonel Ampers hadn't heard my adolescent comment. I wanted him to think that I was more mature than I really was.

"All set, colonel."

"Thank you, Mr. Harris. Let's get on with it, sergeant."

"Yessir."

The driver eased out the clutch and fed fuel to the combustion chamber. The huge truck jerked forward in its lowest gear. By the time the truck reached the abutment, it was rolling along with fits and bounds at about three miles per hour.

"Might I make another suggestion, Colonel?"

"Stop that truck!" the colonel shouted. He turned slightly, and spoke in a calm voice. "Yes, Mr. Harris."

"When a heavy truck starts and stops, the sudden

increase or loss of momentum is transferred to the ground. This has no visible effect on a paved roadway that is well supported, but becomes noticeable on an unconsolidated surface or over a driveway culvert that is buried too shallow. Jamming on the brakes over top of a drainpipe creates such a downward thrust that it has been known to break through macadam and crush the pipe. Now this bridge has been weakened – "

"I see your point, Mr. Harris, and it is well taken. Sergeant, see that the passage is made like Mr. Harris suggested."

"Yessir."

Sergeant Gompton instructed the driver to maintain steady speed once the vehicle was on the bridge.

The driver was a true expert. He backed up the tank truck and started again. Once he attained speed, he kept the truck going straight and smooth and never deviated from his course. A few planks snapped with the retort of a rifle bullet, but most of them merely creaked or groaned. I held my breath as the truck passed over the center pier. I needn't have bothered. The thick wooden balks showed no signs of stress. Thirty seconds later, the truck drove over the eastern abutment onto solid ground.

The collective cheer was stifled by the sergeant's stentorian order. "Let's get that hose truck in position."

Harris jumped into the cab. Quick as a wink, he had the wheels perfectly aligned at the edge of the western abutment. Soldiers commenced immediately to deflate the tires.

Now the Enders brothers approached the colonel. "Yes."

Edward shuffled. "We got a air hose that can fill them tires for you."

Albert added, "If you let us cross next, we can start with the tank truck."

"Good idea. Sergeant – "

"Back it up, Mr. Harris."

"Will do."

While Harris backed up enough to let the pickup

pass on the bridge, Colonel Ampers said, "Thanks for your assistance, Mr. and Mr. Enders. I appreciate your prompt response to my request in Alice Springs."

"Glad to help," the brothers said simultaneously.

"You take the lead once we get all the vehicles on the other side. You're the only ones who know how to find Professor Wayward's camp. Now get moving."

They did. Harris then realigned the hose truck, and crossed expertly without a slipped clutch or change in speed. The troop carriers crossed next; they were light enough that tire deflation was thought to be unnecessary. Then went the volunteers in their off-road vehicles, followed by the two military Land Rovers. The colonel's Land Rover – with me and Sergeant Gompton – was the last to cross.

"Sir, I still can't get through to Woomera."

"Keep trying at fifteen minute intervals."

"Yessir."

The process of reinflating the tires was still in progress. I got out of the Land Rover to watch the proceedings. The Enders had an ingenious device that was useful against off-road hazards: a twenty-foot black hose with a valve depressor at one end and a threaded socket at the other. The threads were located on the outside of the socket. The socket was screwed into a spark plug acceptor after the spark plug was removed. The tires were filled by means of engine compression. The engine ran a little rough because one cylinder wasn't firing, but at idle speed it didn't much matter.

The convoy was off before I knew it, traveling in single file. I tried to pay attention to where we were going, but there were so many forks that all too soon I was lost. The only waypoint I recognized from my two trips with Charles was a wash that had been dry when I first crossed it going, but which had flowing water in it coming back. The water was only a few inches deep, and the hard gravel bottom provided good traction and support for the vehicles, including the tank truck and fire engine.

The sky got marginally brighter with what should

have been first light. But gathering storm clouds blocked most of the predawn brightening. Although we couldn't see it, the sun was above the horizon by the time we reached the Aborigine village.

The place was totally deserted.

The Mold had doubled in size.

Chapter 21

"This is not good," breathed Sergeant Gompton, shaking his head. "Not good at all."

More than half the trees in the orchard were now lying on their sides, where they had fallen after the Mold had consumed their roots. Now the Mold was devouring the bark, limbs, leaves, and fruit. The contamination had spread fairly evenly in all directions through the ground and the sparse understory.

The Mold appeared above ground in fluffy white patches that looked like balls of cotton as large as a bushel and a peck.

"Looks like a bloody blight," added Colonel Ampers. "I want this entire area cordoned off and quarantined. No one approaches the perimeter closer than ten feet."

"Yessir." Sergeant Gompton dashed away to spread the word among the troops and civilian volunteers.

"Does that seem like a safe distance, Mr. Baiker?"

I almost said, "Well, colonel," but I caught myself in time. "Yes, colonel. I think so. Although I would rather have Professor Wayward's advice on the matter."

"You're my expert in Professor Wayward's absence."

This was more responsibility than I wanted to bear, but I didn't seem to have any choice in the matter. "Judging from what I've seen of the Mold so far, the extremity of the organism has always been visible in some form, if you're looking for it. See the slight depression that I was telling you about?"

"I do."

"But I can't say with any certainty that there aren't tendrils extending farther underground – I don't mean deeper in the ground, but extending laterally from the perimeter."

"I understand. Sergeant – "

Gompton was just returning from issuing orders. "Yessir."

"Get someone to dig a hole ten feet from the perimeter. Let's see if we turn up any Mold. And let's not waste any time in spreading our combustibles. I want – "

The trucks were already in motion, and men were starting to surround the area of desolation. There came a shout from the far side of the circle, near the village.

"Let's get over there."

There came more shouts as men arrived at the site of the first shout. The colonel did not swagger as I thought an officer might do in order to convey a sense of decorum. He ran just like anyone else would run in an emergency. I had trouble keeping up with him. Sergeant Gompton was right behind me, huffing and puffing.

A private pulled a writhing form over the slight depression and onto the surrounding dirt. From a distance I couldn't distinguish the shape, although it was vaguely human. As I got closer I thought I saw an anguished female Aborigine wearing a burlap sack over her legs. Upon arrival the reality struck me hard.

Half a dozen men were huddled around a soldier who was pulling the woman along the ground by the arms. He laid her down in a clearing of loose dirt.

"Get awaiy from them!" shouted Colonel Ampers. "Get awaiy!"

The men looked up but failed to comprehend the colonel's meaning. He grabbed a private by the arm and yanked him out of the circle of men, some of whom were standing while others stood in a half crouch.

"Back awaiy! Don't anyone touch her!"

The soldier who had dragged the woman to safety remained on his knees by the injured woman. The woman was screaming like a burn patient.

The colonel resumed his normal intonation. "Darryl, listen to me carefully. We are dealing with an infection that is transmitted by contact. I want you to get away from that woman and doff your uniform.

"Sir?"

"Do it now."

The woman looked familiar to me, but I couldn't quite recognize her features because I was standing at her head, and her contorted face was upside down from my viewpoint. A moment later Albert Enders burst onto the scene with a shout. He stared in horror at the woman whose legs were covered with Mold.

"Patricia! Patricia!"

The woman was so crazed with pain and fear that she didn't respond to his entreaties. Albert started forward, and in that moment I realized who she was: Shane's mother, young Albert's stepmother, and Albert Enders' sister-in-law.

"Grab him!" shouted the colonel.

A private got a partial grip on one arm, but Albert shrugged free and stumbled toward the writhing, screaming figure. Colonel Ampers stepped toward Albert and executed a left-handed, straight-armed karate punch to the solar plexus. Albert exhaled with the sudden pressure of a broken steam valve, and folded in the middle like a snapped twig. He staggered back two steps into his brother's arms, then collapsed, taking Edward down to the ground with him.

With the immediate situation under control, the colonel returned his attention to other important matters at hand. In his normal speaking voice, he said, "Darryl, get out of those clothes now."

Darryl, the private who had dragged Patricia across the ground so heroically, was standing like a statue. Now he came to life like Pygmalion's sculpture. "Yes, sir."

While Darryl stripped, Colonel Ampers turned to another private. "Peter, go get a can of gasoline from the truck. Sergeant, tell Harris to start spraying the perimeter at once. We've got to eradicate every vestige of infection before it spreads farther afield. William, make sure the civilians keep well back from the infected area. We're going to burn everything."

So many events were occurring simultaneously that I could hardly keep them straight. Yet the colonel dealt with everything coolly and calmly, as if he were telling office workers to sharpen pencils and take notes. Albert had not started breathing again. Edward laid him down flat. Patricia stopped screaming, and started choking.

"Sir?" Darryl had stripped down to his skivvies. "Are you going to burn . . . ?"

The colonel knew the first names of every one of his men. "Henry, get that sack of kitchen salt from the back of the Land Rover. Frederick, go find a bucket in the village, and bring it back filled with water. Mortimer, hand me your side arm."

Darryl looked scared. He didn't know if he was going to be shot or burnt, or both. Or in what sequence.

"Darryl, remove the rest of your clothes and put them in the pile with your uniform."

"But sir – "

The troops had no idea of what we were dealing with, or how we had to deal with it. There had been no time to make explanations.

"You'll be okaiy, Darryl. Put your clothes in a heap."

Darryl was visibly shaken as he stripped in the open air.

Edward Enders approached the colonel. "What can we do about Patricia?"

The woman was coughing sputum that was unnaturally dry and white. Her eyes rolled up so far under her lids that her pupils vanished. She no longer appeared conscious. Her chest heaved as she struggled to inhale. White strands of Mold protruded from her flaring nostrils.

"I'll handle it." Without a word to reveal his intentions, Colonel Ampers walked toward the woman but stopped a respectful distance away from her and the Mold that she carried within her body. He raised the pistol with firm resolution, and pumped three slugs into her chest. He stepped back, turned, and handed the pistol to Mortimer butt first. "Reload, and guard this body until we dispose of it."

"Ye-yes, sir."

Darryl now stood completely naked. He was shivering, but it wasn't from the cold.

I was so entranced by the horror of events that I stepped back instinctively, as if I could escape from the gruesome reality. I retreated farther when I saw the fire engine proceeding in my direction. While Harris steered the hose truck, a volunteer firefighter directed the nozzle so that a stream of gasoline sprayed over the perimeter of infection.

In short order, Peter poured gasoline from a jerry can over Patricia' body and along the drag path. He then poured gasoline over Darryl's pile of clothing.

When Frederick arrived with a bucket of water, and Henry with a sack of salt, the colonel waved me over. "Mr. Baiker, mix the salt in the proper proportion in the water."

I gulped. I took a handful of salt and tossed it into the bucket. I stirred the water with my hand, then tasted it. I threw in another pinch for good measure. "That's about right."

"Darryl, wash yourself thoroughly with this salt water. Then I want you to isolaite yourself. Don't move awaiy, and don't let anyone come near you. Peter, go get him some blankets. Don't hand them to Darryl. Toss them from as far awaiy as you can."

"Yes, sir."

By this time the hose truck had nearly completed its circle around the infected area. We had a short respite before the fire engine ground to a halt, and Harris emerged from the cab, and waved.

"Henry, check on Richard and the radio." Richard was the corporal who drove the second Land Rover, which carried the backup radio. "Alert me as soon as he makes contact with Woomera. Frederick, tell Harry to transfer the diesel fuel to the hose truck." Harry was the driver of the tank truck. "Sergeant, do the honors."

Sergeant Gompton pulled a pack of matches from his shirt pocket, swiped one across the striker, and tossed it onto Darryl's pile of clothing. The gasoline ignited immediately. I felt the heat from a distance of twenty feet as flames soared into the air. Fire followed the trail from the clothing pile to Patricia's body, then along the drag path to the perimeter of infection.

With a whoosh that was practically an explosion,

the near part of the perimeter flared up in a huge fire-ball. The intense heat struck me like a bomb blast. I covered my face with my arms as I back away. The fire spread meteorically in a hemispherical wave across the orchard, then sped along the narrow neck of the garden to the foundation of the schoolhouse. The engulfing conflagration was not instantaneous, yet it reached every part of the inundated area in the blink of an eye. An incendiary bomb could not have had a more drastic effect.

A rousing cheer erupted from the civilians, but the soldiers remained stoic. Edward Enders hefted his brother to his feet, and pulled him away from the circular wall of flame. Albert was able to breathe irregularly.

A private named Alistair informed Colonel Ampers that he had made three test digs, and found no sign of Mold in the soil.

"That's good news." To Sergeant Gompton, "Sergeant, gather the men and explain the situation. Make sure the diesel fuel is transferred to the hose truck for mopping up operations. Detail half the men to fan out and crawl along the ground to look for signs of infection. Caution them not to touch anything suspicious – we will simply burn everything that might be contagion. Leave a skeleton crew to watch the perimeter. Dispatch the civilians to search for Professor Wayward and his clan. Their tracks should be easy to follow."

"Yessir." The sergeant ran to issue orders.

The colonel moseyed toward me wearing a deadpan expression on his face. "What do you think, Mr. Baiker? Did we destroy all the Mold?"

I took a deep breath. "I don't see how anything organic could survive a holocaust like this." I glanced around at the gasoline bonfire. Flames nearly licked the low black clouds overhead. "But I've thought before that we killed it all, and every time it came back with a new trick up its tendril."

"Quite."

It started to rain.

Chapter 22

There was little fear that the drizzle would extinguish the blaze. I don't think a full-blown typhoon could have dowsed the flames from the gasoline fire. I found myself wishing for an umbrella. Sergeant Gompton answered my wish with a plastic rain poncho.

I don't know who invented the rain poncho, but whoever it was must have must have been a sadist who invented it to play a joke on soldiers and Boy Scouts. He is probably still snickering in his grave about all the men and boys who have suffered wet feet and trousers throughout the years. The one saving grace was that the temperature was warm despite the rain. In New York, I generally equated getting wet with being cold.

I pulled the rain poncho over my oversized uniform, and adjusted the hood and brim to kept the rain out of my face. The sergeant also gave a poncho to Private Darryl. (I never learned his surname, nor the surnames of any of the other soldiers.) Without protest or complaint, he squatted on the ground and arranged the poncho around the blanket that covered his unclothed body.

The Enders brothers sat disconsolately in their pickup truck.

Colonel Ampers was able to get through to Woomera on the radio. Sporadic static interrupted the conversation, for although he spoke in a low voice that prevented me from hearing his words, I often heard him say loudly, "Please repeat." He did his own share of repeating.

"Reserve troops and equipment are on the waiy," he announced to me and the sergeant. "But this bloody sunspot activity is playing hell with the raidio waves. And the storm isn't helping matters any."

The colonel's second Land Rover stopped next to the troop carriers. I heard a shout of "Hey, Mite," and turned aside to see Professor Wayward, Marie, Charles,

Rosalie, and Arlene striding toward us. They were all wearing raingear that Charles had provided from the seemingly inexhaustible emergency chest that he carried in the back of his Land Rover. Not ponchos, but rain pants and parkas.

Hugs and handshakes were shared all around. Arlene even gave me a lingering kiss – right on the lips, to my pleasant surprise. The professor made introductions.

Colonel Ampers said, "I'm glad to see you alive and well, Warren."

"It feels good to be alive and well, Jonas. We heard gunshots, then saw the flames," explained Professor Wayward. "We started back along the trail and met two men who were searching for us in a Land Rover. They thoughtfully provided us with a lift that was welcome despite the crowding."

"My vehicle ran out of petrol so I had to abandon it," Charles added. "The villagers scattered and took off for the hills. They were the smart ones."

I exchanged looks with the colonel. He said to the professor, "One of them must have returned. We found her body covered with Mold."

"It was Shane's mother," I said.

Arlene gasped and covered her mouth with her hands.

"We incinerated her body along with the orchard and the garden. One of my men maiy have been infected. I've got him under quarantine until we know for certain. Mr. Baiker has been most helpful, but I would like to have your professional advice on how best to proceed."

"You seem to have matters well in hand, Colonel."

"My men are scouring the countryside for noncontiguous areas of infection. What can you tell me about the origin of the Mold?"

The professor hesitated. "Perhaps we had better talk this over in private."

"No need. I have already deputized Mr. Baiker. He is now bound by the Official Secrets Act. I hereby plaice the rest of your outfit under security wraps on their honor. If they will give me their word not to discuss this matter outside of official circles unless I authorize them to do so, I will accept their pledge."

Arlene, Charles, and Rosalie pledged their allegiance at once. Professor Wayward explained the situation to Marie in her language. She replied in kind, and the professor translated her willingness to abide by the colonel's dictate.

"Good. I know it is unusual, perhaps unique, to swear in civilians who have no official standing with either the government or the military, but I want everyone who has been involved in this affair to be cleared, so they can offer advice that is baised on their experience. To do that, they need to know the full extent of the situaition. You maiy speak freely in their presence, Professor."

Professor Wayward nodded in assent. "Thank you, Colonel. These people have been most helpful. Were it not for their active and voluntary participation, I believe the situation might by now be untenable except for measures of extreme severity, which I hope to avoid."

"Quite."

The drizzle slacked so that only occasional droplets of water pattered on our plastic rainwear.

"You should look at this, Colonel." The professor pulled down the zipper of his raincoat, reached into a voluminous pocket of the lab smock that he wore underneath, and brought out a silvery chunk of metal. It measured about six inches in length, was pointed at one end and open at the other. It was roughly cone-shaped but somewhat flattened. It reminded me of a crushed party hat. "Do you recognize it?"

Colonel Ampers hefted the object from one hand to the other. "Quite light. Made of aluminium. The outer surface is scorched." He paused reflectively. "I should say it is the fin tip of a missile." He pronounced the second syllable of "missile" with a long "i," as in smile.

"I agree."

"Where did you find it?"

"In the debris of the schoolhouse after we burned it. It was in the ash in the vicinity of the cabinet where the two boys placed their biology experiment. Did Mr. Baker tell you how the Mold started and spread?"

"He did an exemplary job of describing every aspect of the contagion in exacting detail."

"Good. Then I will proceed directly with some educated speculation. I made the same identification that you did. I believe that this piece of metal is from the missile that went missing last month. I believe the boys found it during their amblings around the countryside. I believe that something was growing inside the cavity: perhaps an exotic microorganism which has lain dormant in the upper stratosphere, which was acquired during the flight, and which, in the presence of a source of nutrition and extracts from wild local herbs, infected the mold starter that the boys used to initiate their experiment. I believe that the happenstance of combined organic compounds caused the original organism to metamorphose into a fast-functioning metabolic monstrosity whose only imperatives are growth, expansion, and propagation.

"This Mold is not a simple infection. It is the vanguard of an invasion from the fringe of outer space."

Chapter 23

The professor's pronouncement came as a revelation to me. For the first time since the dramatic appearance of the Mold, I fully comprehended the malignant character of the abomination that we were dealing with. The implications were enormous. Poised as the human race was – on the verge of breaking free of Earth's gravitational attraction, with the intention of expanding exploratory horizons throughout the solar system – the discovery of inimical life forms at our very doorstep boded ill for these enterprising ventures.

Only three years earlier, before his tragic assassination, President Kennedy had made a vow for America to send a man to the Moon and return him safely to the Earth by the end of the decade. If the upper stratosphere was suffused with opportunistic biological entities, every spaceship that ventured outside the Earth's protective envelope possessed the potential to return with a cargo of organisms that were lethal to life that occupied the surface environment.

The possibilities were appalling. The fate of the world lay literally at our feet.

Colonel Ampers wasted no time in dispersing his troops to search for the crash site of the errant missile. The consensus of opinion was that Shane and Albert had discovered the site during a recent weekend sojourn. The professor doubted that the boys had stayed out overnight. This implied that they had found the rocket fin on a single-day trek; or, no more than half a day's walk from the village. But Aborigines were fast walkers, and they could travel a long way between sunrise and noon.

The civilian volunteers joined in the search. One of them took Charles and a jerry can of gasoline to the place where he had been forced to abandon his Land Rover. The colonel wanted every available vehicle involved in the search. There was no telling when the reinforcements from Woomera would arrive.

Rosalie busied herself by wandering through the village, partly to look for anything out of the ordinary, and partly to search for stray pets that might have been accidentally left behind in the mad exodus to escape from the reach of the Mold. Marie rustled some food from the kitchen. Professor Wayward, Arlene, and I took refuge in the professor's dining room, seated around the table. I was so tired that I could hardly keep my eyes open. None of us had slept the night before. But Marie's hastily prepared victuals perked me up and renewed my energy.

Professor Wayward waxed speculatively about the life cycle of the Mold. "We must examine every possibility, but I do not believe that there will be any signs of infection at the crash site. I believe that this extraordinary Mold did not metamorphose to its present morphology until the extraterrestrial entity was mixed with Earthly organic compounds in an unlikely combination."

"But the raw material might still exist," I disagreed.

"Yes," Arlene chimed in. "The original organism – microbe or bacterium or whatever – might be able to adapt in other ways. Maybe in ways that are worse – if that is possible."

The professor shrugged. "Anything is possible when dealing with biogenesis. The sum total of our biophysical experience is with organisms that inhabit our planet's lower reaches: under water, on land, and in the air. We have no idea what organisms we may encounter outside of our protected environment, where the harshness of nearby space may affect evolutionary processes. Now that Jonas Ampers has bound you to the Official Secrets Act, I am permitted to impart information that I was previously forced to withhold."

"You mean, about the experiments that were conducted at Woomera?" I asked. "The atomic bomb tests?"

Arlene gasped.

Before she could ask, I said, "I'll fill you in later on what Colonel Ampers told me."

"Not about nuclear testing per se, as that was not within my scientific purview. I was tasked with the responsibility of studying the effects of hard radiation on organisms that survived the detonations: those that surrounded the blast area and were not killed outright by the explosion. Of course, for a great while scientists have been conducting laboratory experiments in a similar vein, such as by irradiating insects and observing the genetic mutations that occurred as a result."

"Drosophila melanogaster," I said.

Arlene looked at me as if I had horns sprouting from my forehead. I merely shrugged.

"Quite right, Mr. Baker. The common fruit fly was the most frequently used test subject because irradiation caused pronounced mutagenic aberrations in succeeding generations. Instead of laboratory experiments, I did actual fieldwork – and on all biota. I ascertained very quickly that most of the flora and fauna that were subjected to nuclear fallout died within a very short time after extreme exposure. Those few that survived and were able to procreate produced mutations that generally failed to achieve adulthood. In the very rare cases in which offspring reached the age of reproduction, the mutated genes were not transmitted faithfully because the genetic alterations were transient. The resulting monsters – teratogenetic carriers – never stabilized. Inevitably they died. In the rare instances in which there *was* a succeeding generation, additional mutations occurred and always for the worse, resulting in premature termination of the resultant offspring. The end result was always the same: mutated individuals died in either the second or third generation, and failed to produce a fourth."

"So are you saying that atomic bomb radiation never spawned a radically new species?"

"Correct. Ironic as it may sound, the tests of warfare did not ultimately lead to viable deformities. Yet the tests of peaceful rockets that were intended for the pursuit of scientific knowledge led to the capture of organisms which possessed the ability to alter the planetary environment, perhaps permanently."

"Do you mean the Mold?" Arlene asked.

"Or the organisms that the government destroyed at Woomera by the detonation of atomic bombs?"

Once again Arlene gave me a look that could turn flesh to stone.

I said, "Later."

"I mean the Mold. That previous organisms I was not permitted to study. Nor did I desire to do so. Their mere existence was too grave a threat to life on the planet. I recommended total destruction before the unknown and unpredictable infections could become widespread. The government sanitized the area with, uh, extreme severity."

A light went on in Arlene's eyes. She had heard that expression before. "Do you mean to say that if we don't eradicate this Mold before it proliferates, the Australian government will sanction its destruction by means of – a nuclear device?"

"Precisely, Miss Hawkins."

I added, "Tough times call for tough measures."

"But – " Arlene sputtered, but didn't know what else to say on the spur of the moment.

"Mr. Baker is right. Extreme severity may be necessary in order to ensure the survival of humanity. I believe that this incipient life form that we call Mold is not a simple mutation that will terminate after two or three generations. Indeed, it may not reproduce at all, but continue to grow exponentially. I greatly fear that it will persist in perpetuity if we do not eradicate it forthwith. The mutagenic effects of nuclear fallout are well established. But we are only just beginning to explore the reaches beyond Earth's atmospheric envelope.

"Out there, raw organic material is exposed to intense ultraviolet light, cosmic radiation, alpha and beta particle bombardment, broad temperature variations, magnetic influences, gravitational stresses, energy fluxes, ionic potentials, and who knows what other forces that mankind has not yet discovered. I hypothesize that these upper stratospheric carbohydrates constantly undergo multitudinous changes in molecular structure as a result of these various hostile agencies.

"Generally speaking, these organic molecules pose no threat to the Earthly environment because they cannot interact with tellurian – that is, terrestrial – organisms. They exist thirty miles above the ground. Those airborne molecules that manage to escape naturally into the troposphere may take months or years to reach the surface, by which time their structures have achieved a state of harmless equilibrium. But when we bring them back quickly, trapped in the interstices of missiles or rockets, they are still in a solvent or chemically active state. That is why strict decontamination procedures are so essential for returning spacecraft. Unfortunately, my opinion of the matter is generally ignored in the interests of expedience and economy.

"Exobiology – the study of life that originated elsewhere than on Earth – is a brand new field about which we know practically nothing. Only now are we on the threshold of learning the dire side of otherworldly life forms.

"Think about this: viruses, as you know, are inimical to our way of life. A virus is a parasitic molecular chain consisting of naked ribonucleic or deoxyribonucleic acid that is surrounded by a protein sheath. A virus must invade a host cell in order to replicate. Thousands of viruses are known to exist, in every ecosystem on the planet. Scientists have been studying viruses for more than half a century. Yet we cannot cure even the common cold or influenza virus.

"Now imagine a molecule that can infect Earthly organisms in a similar fashion, but a molecule that does not rely upon nucleic acids for replication. Its biology is totally alien. It carries its genetic information in some manner that is unknown to us. Most exobiological molecules must be so vastly different from our own form of life – which relies upon RNA or DNA for its very existence, much less its reproduction – that they cannot interact chemically with living cells.

"But because billions of these exobiological molecules might inhabit the upper stratosphere, created by cosmic forces in never-ending variety, every once in a while one of them happens to possess the blind ability to combine with our form of life. I believe that that is what we have encountered here: a chance miscegenation, as it were, between our form of life and an otherworldly form of life that has produced a viable mutation."

I sipped my tea in the quiet that followed the professor's theoretical abstractions. Arlene did the same. Sitting around the professor's dining room table was beginning to feel familiar, almost like being home. With the Mold now beaten, I should have been asking questions that pertained to my article. But the Mold and the potential consequences of space exploration had redirected my thoughts. Whereas before I was eager to sound out the professor's scientific views, now I was fearful of what forbidding predictions he had to make.

Finally I cleared my throat. "I suppose we can be thankful that the Mold isn't mobile. That this otherworldly organism infected a stationary fungus instead of a flying insect."

Arlene shook her head. "What worries me is that otherworldly infections have occurred three times so far and we've only been in the space age for seven years."

I nodded. "*Sputnik*, October 4, 1957."

Arlene grimaced at me.

I shrugged visibly. After all, as a science reporter, it was my job to know science. I voiced a stray thought, "A horde of ordinary locusts can lay waste to entire lands. Imagine the damage that infected locusts could do." A way-out scenario suddenly materialized in my mind. "Hey, do you suppose that a contamination like this could have killed off the dinosaurs at the end of the Cretaceous? I mean, more than fifty percent of all the existing species on the planet went extinct practically overnight."

The professor nodded. "It is certainly a possibility."

To Arlene, "The most commonly accepted theory for the demise of the dinosaurs is a big comet that struck the Earth and threw such huge clouds of dust into the air that the sun was blotted out for years and years. No sunlight means no photosynthesis. First the plants die out, then the animals that feed on plants, then the animals that feed on other animals. A natural domino effect."

Then another startling thought manifested itself. "And maybe, instead of a big comet that exploded on the ground, a small comet or some other astronomical body brought an alien organism through the atmosphere – shielded it so it didn't burn up from friction."

"That is also a possibility."

That last flash of insight seemed to have sapped my strength. Suddenly I was so sleepy that once again I had trouble keeping my eyes open. The food and tea hadn't invigorated me as much as I thought they would. With the sun about to set, I realized that I had been awake for some thirty-six hours – and had run a marathon in the middle of that period of wakefulness. I propped my chin on my fist, with my elbow on the table. I found myself without additional insights.

Arlene picked up the ball. "You just said something about exponential growth. Yesterday – or some other day – you mentioned food supply. Now, from what little I remember from Biology 101, nothing can live and grow without a source of energy. As Tim said, plants get their energy from the sun. You said that molds and funguses get their energy by dissolving organic material. It seems to me that this Mold can't grow forever because it won't be able to feed itself. I mean, the middle of itself. . . . "

She knew what she wanted to say, but not how to say it.

The professor understood. "I have been pondering that very point. Trees rely upon osmosis to transport water from their roots to their leaves – "

"Osmosis?"

"A mechanism by which diffusion of a liquid is accomplished by passing the liquid through a semipermeable membrane from one cell to another until the concentration of liquid is equalized in both cells."

"Oh, yes. I remember."

"Plant growth is fast at first because liquid nutrients do not have far to travel. As a plant grows larger, the growth rate is reduced because of the increasing distance that nutrients must be transported. This transportation rate is a function of physics, not biology. Ergo, at some point the Mold will have spread itself so far outward that it will be unable to transport nutrients to the center of its anatomy."

The professor pulled a tablet from the voluminous folds of a lab coat pocket. "I did some preliminary calculations based upon the approximate time that transpired as we watched the growth ring expand by several inches. Unless the Mold possesses some mechanism for siphoning nutrients from the air – "

"Can it do that?"

The professor looked at her in perplexity. "I do not know. Earthly plants do not operate in such a fashion, but we are dealing here with an unknown quantity, so I thought it best to take all possible factors into account. Stationary animals that live under water – corals, barnacles, and sea anemones, for example – survive by capturing nutrients in the water that flows past their tentacles. I considered the possibility that this alien Mold might capture airborne particles, such as pollen, as an additional means of supporting its increasing body mass.

"For what it is worth, my calculations lead me to suspect that the Mold could be dying from the inside out."

"That's a relief," I managed to mutter.

"Undoubtedly its root tendrils extend down into the ground until they run out of organic material or reach bedrock. However, this is not necessarily meaningful, as the outer perimeter will continue to expand and consume everything it encounters, leaving sterile ground behind as the core is starved of nutrients. At this point in our knowledge of the Mold, I suppose we could expect anything.

"Anything?" The tone of Arlene's voice was one of despair, or surrender. I don't want to say again that she was wishy-washy, but her character seemed to vacillate constantly between that of a tough infallible tomboy to that of a stereotypic fragile female. I suspected that she was more complicated than either extreme would lead me to believe.

Her initial resentment of me had slowly yielded to acceptance. Her initial pride had given over to – dare I say it? – passion. Her strength now seemed to be wavering. But I suspected that this latest exhibition of weakness was ephemeral, accentuated perhaps by exhaustion.

"Anything," repeated the professor. "Including the unexpected."

Chapter 24

"Hiy, Mite!"

My last conscious memories consisted of a dreamy montage: my chin falling off my fist and nearly banging against the table; Arlene putting her arms around my shoulders; the professor saying, "Oh dear;" Arlene saying, "You can sleep in the bed;" stumbling into the bedroom; falling onto the mattress; feeling a blanket being tucked around my body; my shoes being pulled off . . .

I managed to force one eye open partway. I squinted in the dim light. Charles' infectious grin stared down at me from the middle of the doorframe. Something tightened across my chest. The constriction was soft and warm. I thought immediately of a python, and in my sleep-slurred mind I tried to remember if that species of snake lived in Africa or Australia. Abruptly I recalled the answer: both!

I hardly dared to breathe. Then the felt a strong pressure against my sternum, followed by movement behind my back.

Arlene pushed herself up onto her elbow, and dislocated the blankets in the process. "Oh, hi, Charles. Is it morning already?"

"Well past."

Arlene was snuggled against my back like one spoon snuggled against another. She removed her arm from around my chest so she could brush back her hair. "Have you been up all night?"

"I catnapped in the Land Rover, then spent the early morning hours in a hut."

I was trapped in the blankets. I struggled to rise. I sneaked one hand out from under the covers, then the other. "Just barge right on in."

Arlene rolled backward to give me room to stretch. I was immediately sorry to lose contact with her body.

"I couldn't let you sleep the diy awiy."

"Yes you could." By this time I was regaining consciousness. "I could use a few more winks."

"Couldn't we all, Mite."

"I know I could." Arlene sat up but stayed close enough so that her hip touched mine. Her long hair was disheveled, looking much like a mop that had dried and stiffened in a bunch. I didn't mention my observation.

I stretched my arms carefully so as not to punch Arlene in the face. Yawning, I slurred, "The dawn of a new day. Or is it?"

"It's a new diy but without the dawn. The rine has slacked off, but the forecast is calling for the storm to get worse before it gets better." After a pause, "How would you like to go for a ride?"

"Where to?"

"The rocket crash site."

Arlene let out a yelp. "They found it?"

Charles jerked his head. "Spread over a dozen acres of desert. The soldiers are collecting the parts."

Speaking of soldiers, "Hey, what about, uh, what's his name? Private Darryl?"

"Clean bill of health."

"Good thing they doused him with saltwater," said Arlene.

"Gasoline would have worked just as well."

She punched me playfully on the shoulder. "Oh, you. How can you joke about a thing like that?"

I punched her back. "It's my nature. It helps to relieve stress."

"If you two can quit bantering, Marie has packed a lunch that you can eat along the wiy."

"Lunch," I complained. "What about breakfast?"

"Too late for that, Mite." Charles turned and strode out of the room.

I glanced at my watch through still-blurry eyes. It had stopped running. I shook my wrist and held the watch to my ear. It ticked twice. I had forgotten to wind it.

"Ten forty-three," Arlene announced. She held out her wrist for me to see.

"Brunch." I swung my aching legs out of bed. The

nighttime marathon had caught up with me. My calf and thigh muscles were sore from running such a distance, and from running on a sandy roadbed. I eased my feet into my shoes, laced them, and stood up on shaky legs. "I guess I'm going to miss the New York Post Collegiate Marathon."

Arlene was already dressed. "Tim, you won a race that is far more important."

I groaned as I bent my shoulders back. "I guess you're right." I pointed to the bed. "Does this mean that you're no longer mad at me?"

"I was never mad at you. I was just mad."

"Mad as in crazy – ?"

"Hiy, if you two can quit bantering . . . "

"All right already. I'm coming."

"You're not even breathing hard."

Now it was my turn to stare at Arlene as if she had antennae waving over her ears. She didn't give me an opportunity to respond.

With a strange smirk on her face, she grabbed me by the hand and pulled me along after her. "Come on, Tim. Let's go."

The professor and Rosalie were already sitting in the rear seat. Arlene squeezed in beside him, while I rode shotgun. I wanted to ask questions, but Charles handed me a sandwich, and I found that my hunger for food for my stomach was greater than my hunger for food for thought. I started munching.

The Land Rover jounced over hill and dale. Charles swerved often to avoid the worst of the obstructions. The constant turning and up and down motion made me slightly car sick, but I was so hungry and so distracted by dire thoughts of what we might find at the crash site, that I managed to overcome my nausea.

I barely had time to finish my sandwich before Charles halted atop a rise that overlooked a rugged, boulder-strewn stretch of land that filled a shallow depression that was bisected by a dry riverbed. Except for the black bulbous clouds that brooded on the horizon, it could have been a moonscape.

"How desolate, but how beautiful," Arlene said.

"This is our land," Rosalie commented quietly.

"A sight you don't see in New York City, right, Mite?"

I was absorbed by the stark grandeur. "Not since the Pleistocene."

Charles eased out the clutch, applied some gas, and let the Land Rover roll down the gentle incline. Soft sand nearly brought the vehicle to a halt. He depressed the gas pedal. Sand spit out from under all four tires, and the Land Rover did what it did best by moving forward through terrain that would have balked most vehicles.

The Land Rover churned across the desert with the ease of a Volkswagen plowing over thin snow on Broadway. Although it lay only a mile away, it took us ten minutes to reach the dry riverbed. Charles then proceeded "upstream" for about fifteen minutes to a plateau that looked like the parking lot at Macy's during a one-day sale: vehicles faced in all directions with no two aligned in parallel. Only the tank truck and fire engine were missing.

Charles ground to a halt at the makeshift command post that consisted of both army Land Rovers. Sergeant Gompton was talking on the radio. Colonel Ampers and Professor Wayward were examining missile parts in a large pile of debris, while soldiers and civilian volunteers arrived in a continuous stream with shards and pieces of metal.

We all piled out of the Land Rover.

I decided to follow the colonel's previous advice and get right to the subject at hand, with no "well's" or "uh's." As soon as my feet hit the ground, I asked, "How did you find the crash site so fast? There must be hundreds of square miles where the rocket could have fallen, radiating from the village in any one of three hundred sixty degrees."

Colonel Ampers grinned. "A combination of fortuitous circumstances and deductive reasoning."

The professor pointed to Rosalie. "Rosalie reminded

me that the villagers spotted a falling star about a month ago. She informed me of the quadrant in which the streak appeared across the nighttime sky. That must be what started Shane and Albert on their trek. I remembered the rumor that Charles repeated about a missile going astray, and the colonel confirmed the time frame. We fanned out in this direction until we stumbled over it."

"Scientific method in action." I was impressed. "Along with some anecdotal observation that scientists deplore."

"Quite," said the colonel.

"What's the diagnosis?" Arlene asked.

"Nothing so far."

The professor spread his hands. "There is no sign of contamination of any kind. Whatever chemical reagent or quasi-organism was brought back from the troposphere has either become inert, or it was never activated by admixture with Earthly herbal ingredients."

"And there are no other forms of – contamination?" I queried.

"Not that we have seen."

"So the Mold is dead," Arlene asserted hopefully.

"As far as we can determine."

The relief that washed over me felt like cool water after a steam bath. "This calls for a celebration. Maybe Marie could spike some tea for us."

"I wouldn't like to be premature," started the colonel. "We still have a crash site to clean up." He swept his arm to indicate that people who were collecting rocket parts. Many of them were crawling on their hands and knees. "Not until we examine every last piece can we be certain that there is no secondary contamination. You people go back to the house. My men can taike caire of this."

"What about the reinforcements from Woomera?"

"They are still on the waiy."

"That seems like overkill," Arlene ventured.

"Perhaps the measure seems excessive, but I won't call them off until we are one hundred percent certain that the situaitaion is secure."

I nodded knowingly. "A wise decision."

"Right on," said Charles.

The decision was wiser than I could have imagined.

Chapter 25

The view from the professor's house was bleak and barren. The schoolhouse was gone, the garden and orchard were burned out cinders, and the coal-black sky was threatening a downpour at any moment. Light drizzle portended the approaching storm.

There was no electricity because there was no fuel for the generator. The well pump that served the house couldn't run without power, and the dining room was dark despite the mid-afternoon hour. Rosalie took a bucket to fill with water from the hand pump in the village, while Marie served tea that she heated on a propane stove.

"I can understand why Colonel Ampers wants more men on the job," Arlene opined, "but I don't see why he has to have tanks and artillery pieces."

I snickered. "The only difference between men and boys is the size of their toys."

"How big is your toy," Arlene shot back.

My jaw dropped open. I was speechless at her open audacity.

Charles guffawed. "She got you there, Mite."

Even Professor Wayward cracked a smile at my expense.

Arlene winked at me.

Someday I'm going to have to spank that girl.

Someday. But not today.

The professor donned a serious mien. "Colonel Ampers is simply following military protocol. When dealing with an enemy of unknown size, uh – " He grimaced at his unintended faux pas. " - and, er, strength, one must muster all the forces at one's command. A brigade of tanks and artillery may seem like an excessive show of force, but remember that we are dealing with a quantity whose resourcefulness has proven to be extraordinary. It is better to have reinforcements that are not employed, than to need reinforcements that were never called."

"Right on."

I agreed with Charles and went one better. "Not only that, but the Mold has reacted in ways that were totally unexpected – at least, for an Earthly mold. As you noted – " I nodded to the professor. " – molds grow in the darnedest places and under the darnedest conditions."

Arlene interjected, "But no mold, or any other living creature for that matter, can grow without a continuous source of nourishment. Once the food runs out, the creature must die – "

"Or go into a state of dormancy," inserted the professor.

I shivered. "Don't even think about it."

The professor shrugged. "Seeds that were found in Egyptian tombs were able to germinate after being dormant for two thousand years."

"Maybe so. But those were seeds that were intentionally preserved, not a Mold that has been eradicated on three separate occasions."

He shrugged again.

"As I was saying, this Mold did far more than respond to its environment. I've been thinking – "

"A drum roll please."

I glared at Arlene's obvious innuendo, wondering what she had put in her tea to make her so boisterous. "As I was saying, this Mold seems to have *anticipated* potential changes in its environment. Almost as if it *knew* what it was doing."

"Is this another one of your stories?" Arlene said accusingly.

"No, hear me out. Stimulus and response are standard biological imperatives. Poke a flatworm in the side, and it moves away from the point. Pour acid into a drop of water, and an amoeba will scamper away from the source of irritation. Put your hand on a hotplate and you will lift it pretty darn fast. But if you'd never seen a hotplate before, and had no idea that it could burn you, or freeze you, or dissolve your skin, how could you possibly prepare yourself for every harmful eventuality?"

Three blank stares appealed for me to continue.

"Okay, let me give you an analogy. Lab tests have proven that DDT kills mosquitoes. But when you spray a swamp, you come to learn that DDT doesn't kill *all* the mosquitoes. Certain individuals are resistant to DDT. These survivors breed a new strain of mosquito that passes on the resistance to the next generation. Eventually you have just as many mosquitoes as you started out with, because more DDT treatments don't affect the new and improved strain. Now you have to discover a new poison to kill the strain that is resistant to DDT. And so on, and so on. You can keep down the population by continuously trying new poisons, but there are always survivors. That's because mosquitoes – all insects, for that matter – not only have an incredibly large gene pool, but are readily subject to mutation.

"Now take our Mold. According to our hypothetical evaluation, an ordinary Earthly mold has been imbued with something from outer space: a biological organism, or a viral infection, or a raw chemical chain that's equivalent to DNA – something that has altered, or mutated, the mold's chemical structure in such a way that it has become a super Mold with incredible capabilities; a Mold that doesn't just *respond* to stimuli, but one that foresees a multitude of probable stimuli.

"You said as much yourself, professor: this Mold isn't necessarily a single organism, but perhaps a colony of cooperative strands that can survive on their own if they are separated from the main mass. Remember that we are dealing with a Mold that contains an Earthly component complete with an evolutionary history, or racial memory, and an outer space component with unimaginable functions. This miscegenation may imply either a huge genetic pool, or the inborn ability to mutate in order to meet adverse circumstances and conditions, or both. Ultimate survivability."

Professor Wayward cogitated.

Arlene stared at the ceiling. "You make it sound possible."

"Wait. There's more.

Charles snorted. "I'll bet it gets worse."

I held up my hands, palms outward. "Hear me out. On my previous assignment, I interviewed Dr. Wolfgang Brunholf, a brain physicist who conducts studies on cerebral development and human intelligence – as well as mentality in all its forms, in animals and plants."

"Vegetable intelligence!" Arlene shouted. "Now I know this is another one of your stories."

I held up my hands again, and repeated, "Hear me out. Please." When the shuffling and hard stares ceased, I continued. "The definition of intelligence is a human construct. Dr. Brunholf surmises that there are many kinds of intelligence, and that most of them are not understandable to us because of the way our brains are wired. What we think of as intelligence is related to how we perceive the world around us: through our senses. Animals with different senses, or with heightened senses, perceive the world differently. For example, we rely on sight while dogs rely on smell. Dogs smell the world before they see it, and their concept of the world differs as a result of their primary perception. Dr. Brunholf's point is that dogs are not necessarily less intelligent than human beings, but that they respond to different stimuli. If all animals were suddenly struck blind, canines would rule the world because they could smell their way around it.

"Dr. Brunholf also surmises that what we call instinct is in reality a different form of intelligence. Or many different forms. We label it but we can't understand it, just like we label infrared and ultraviolet even though our eyes can't see those color ranges, or any other wavelength in the electromagnetic spectrum outside of red, orange, yellow, green, blue, indigo, and violet. We know that infra and ultra wavelengths exist because we have instruments that can detect them. By the same token, Dr. Brunholf postulates that forms of intelligence exist that we can't comprehend."

Arlene was agnostic. "Great. Now we have a smart Mold that can outwit us at every turn."

"I'm not saying that, but I *am* saying that the Mold has responded in ways that give the appearance of intelligence."

Charles was less noncommittal. "You mean, like when we threw saltwater on the Mold the first time, and it went and hid and grew undetected inside the walls and ceiling of the schoolhouse?"

"Precisely. And after we destroyed it the second time, it spread through the ground where we wouldn't think to look for it."

Charles winced. "It does seem right smart when you look at it that way."

"I'm not saying that the Mold can formulate ideas, or comprehend its situation, or has cognitive functions and imagination, or that it is self-aware. I'm only saying that it has responded to its environment and to possible environmental changes in ways that it appears to be making conscious decisions on how best to survive."

"Now we are confusing biology with philosophy," said the professor.

It was my turn to shrug. "To a certain extent I agree. Nonetheless, the Mold has exhibited an astonishing resilience to being snuffed out, by saving bits of itself and then – dare I say it – learning from the experience."

The professor raised first one eyebrow, then the other.

"Learning implies intelligence. A kind of intelligence that's different from human intelligence, but intelligence just the same."

Charles quoted Shakespeare. "A rose by any other name would smell as sweet."

Now Arlene shivered. "That's a creepy idea."

The professor ruminated silently.

"I don't believe that the Mold can actually perceive us, or conceptualize danger, but it seems to me that it possesses an uncanny ability to pre-empt all attempts to exterminate it." I shrugged. "It's just food for thought."

Arlene humphed. "Our thoughts, or the Mold's?"

"Take your pick." *Or shovel*, I added mentally, but

kept my mouth shut because the moment was too serious for juvenile humor.

The professor was still ruminating in the subsequent silence, when Rosalie entered the house without the bucket.

"Come see, Professor."

Her words gave me more than the heebie-jeebies. I remembered what happened the last time Rosalie made such a request. I panicked. My spine tingled as if I had been shocked by an electric charge.

We all rose at once.

The storm was in full swing by now, and rain was falling hard. Cold water droplets added physical chill to the emotional chill that was far more debilitating.

Rosalie pointed to a scattering of white dust that powdered the dry ground under the kitchen window: a spot that was protected from the pelting rain. It looked innocuous, like spilled talcum or shaved chalk, or perhaps grains of pollen.

"Charles, would you mind retrieving my microscope from the pile of debris?"

"It's half melted, Doc Warren."

"Please."

Charles nodded and did as the professor bid.

The plastic base of the microscope was a molten blob. The eyepiece support assembly was distorted, like a caricature from the bizarre imagination of Salvador Dali. The professor gripped the two eyepieces and twisted them apart. The left eyepiece broke off the main mass. He tossed the remains of the microscope away, then got down on his hands and knees over the white dust.

He put the eyepiece to his right eye and adjusted the focus. He examined the dust long and hard, then longer and harder. Finally he leaned back on his knees and toes in an attitude of prayer.

"Mold spores."

Interlude 7

The Mold sensed the presence of food. Food lay all around it: in the ground below, in the surrounding earth, in the towering structures above.

The Mold also sensed the presence of imminent danger. The danger was above, but it saturated the soil as it filtered through the granules that constituted the medium through which the Mold extended its growing strands.

Deep within its cell structure, encoded on chains of nucleic acid, was information that enabled the Mold to choose various courses of action in response to external stimuli. Merged with this encoded information were new and aberrant codes, alien codes, constantly changing codes.

These codes triggered a multitude of responses. Most of these responses were ineffective; some were marginally advantageous; a few were spectacularly efficacious.

Instead of using its nutrient supply to continue its growth, the Mold switched modes to reproduction. There were numerous ways in which the Mold could reproduce. It tried them all.

They way that worked best and ultimately survived the holocaust was sporogenesis.

*   *   *

Marie hummed to herself as she worked in the kitchen in front of the open window. Everyone else was gone. The villagers had evacuated to the distant hills; the civilians and soldiers had gone to explore the countryside; Charles had taken the professor, Rosalie, Tim, and Arlene in his Land Rover.

She was alone in the village.

The desert air smelled fresh and clean. The recent rain had washed the impurities out of the atmosphere. She inhaled deeply during the lull between storms, filling her lungs to capacity.

Black clouds obscured the horizon, the harbinger of

another storm. Marie watched the storm's approach with enthrallment. She was a desert dweller, and loved the many moods of the land.

Gradually the air turned musty. She coughed as a mote tickled the back of her throat. The musty odor grew stronger. She coughed some more. And some more.

Then the rain returned and the air cleared itself. She drank a glass of water, and felt much better.

Marie hummed again. She smiled when she saw the professor returning. Quickly she prepared some tea and a small platter of biscuits and cheese. As the professor and his visitors ate noisily in the dining room, she went about her household tasks.

Then her cough returned. Her stomach felt queasy and something was lodged in her throat. Her skin felt itchy. She scratched her arms, then her legs, then the orifices between her legs. She coughed hard, and in the process dislodged a lump of what appeared to be cotton. The fluffy pellet spat into the sink, followed by another that was the size of a golf ball.

She stared at the white puffy sphere that looked like a cat's fur ball. Instead of a smooth outer shell, the surface of the sphere was covered with tiny hairs that waved and writhed like a nest of baby snakes. She was afraid to touch it.

Then she noticed the tiny white hairs that were growing out of her coal-black skin. Her arms and legs were coated with fluttering down.

Her throat became congested. She found it difficult to breathe. Her mouth tingled. She stuck her fingers between her lips, and scraped the fluff that coated her pink tongue.

She coughed again and cleared her throat.

Then she screamed.

Chapter 26

The gurgling scream that pierced the air was muffled. It imparted a tone of fear that I had never heard before, yet I knew its meaning nonetheless. My spine cringed and crinkled with colliding chills.

I saw Marie's fuzz-covered face at the kitchen window. She looked as if she had dusted her black skin with white baby powder. Her mouth was open, but it was filled with white fluff, as if she had tried to swallow too much cotton candy. She fell backward out of sight with a clatter of dishes.

Charles reacted instantly and beat me to the door. He stopped at the kitchen doorway with his hands on the jambs. I didn't try to get past him.

Marie was standing with her back against the wooden countertop. Her eyeballs were the size of quarters and were surrounded by a fringe of white filaments. Her head bobbed irregularly; her arms twitched. She looked as if she were wearing a fur coat made from the skin of an albino wolf. Her chest heaved as she tried to breathe through the plug that filled her mouth.

The rasping gurgle that escaped her lips was horrible to hear. It reminded me of a diesel engine that was badly out of tune. She was choking to death, and there was nothing that we could do about it.

Her body was suffused with Mold.

Slowly, as her legs gave way, she slunk down to the floor, where her body was consumed from the inside out.

I heard a gasp. When I turned my head, I saw Arlene holding her hand to her mouth, and the professor with a dazed expression like a deer caught in headlights. It was a replay of the scene with Shane.

"What do we do?" Charles shouted.

"Burn down the house," Arlene replied. "Quick! Before the Mold spreads."

"What about us? If the spores are airborne . . . "

I was thinking hard. "Let's hope that the rain beat them down. Marie must have inhaled the spores during the lull in the rain, while we were at the crash site."

"I hope you're right."

"So do I. Professor . . . "

It was no use talking to him; he was paralyzed. Arlene and I grabbed him by the arms and led him out of the house; he walked stiff-legged as if he were wearing stilts. Charles was right behind us.

"What do we do now?" Arlene wanted to know. "Siphon more gasoline from the Land Rover?"

Charles nodded toward the tank truck and fire engine. "There's petrol and diesel fuel over there. There should be dregs in the storage tanks, and if not, there's some in the fuel tanks. I'll go check."

"Marie . . . "

"She's dead, Professor, and there's nothing we can do about it." I glanced around for a place to sit him down. I dragged him to the generator shed with Arlene in tow, and pushed him down onto a makeshift bench. "We've got to incinerate everything, and hope that we haven't inhaled any spores."

Arlene was horror-stricken. "How can we be sure?"

I glanced at my watch and made a quick calculation. "We should know in about two hours."

"My god."

"Mine too."

I heard the roar of a diesel engine. Charles was in the driver's seat of the hose truck, spinning the steering wheel as the heavy-duty vehicle lurched forward. He ground to a halt in front of the house and leaped out of the cab like his pants were on fire. I ran to help him man the equipment.

"Should be a leftover mixture of petrol and diesel. We'll give it a squirt, Mite, and if that don't work we'll disconnect the feed line to the fuel tank."

"Sounds like a plan."

I dragged the hose off the reel while Charles fiddled with the pressure pump. In short order the pump was forcing gasoline out of the bulky nozzle. I wasn't prepared for the sudden burst of pressure. The hose whipped like a snake and yanked the nozzle out of my hands. Precious gasoline and diesel fuel sprayed across the ground until Charles helped me catch the wildly gyrating nozzle. We held onto it together and played a stream of inflammable liquid against the side of the house and through the open window into the kitchen.

We then hauled the hose around the corner of the house in order to inundate the adjacent side. The flow was quickly reduced to a trickle, but not before we sprayed the mixture through the outside doorway.

"Think that'll be enough?" I wondered.

"It'll have to do."

We dragged the hose away from the house. Arlene was already standing next to the fire engine with a broom that she had ignited and held like a burning brand.

"I'll do it." Professor Wayward trudged past Arlene and pulled the torch from her grasp. "I need to do it."

Despite his pronouncement, he stood for a long time and contemplated his actions. He contemplated for so long that I was getting anxious. I was about to snatch the torch out of his hand and do the job myself, when he took a final step forward, bent at the waist, swung his arm behind him, and tossed the flaming brand underhanded like a softball pitcher in the ninth inning of a game with two players out.

The burning broom flew right through the kitchen window. The house burst into flames with a gigantic whoosh and a blast of heat that made us all back away. Although the rooms in the rear had not been primed with incendiary liquid, the fire quickly spread until the entire building was engulfed in flames: a funeral pyre for the professor's faithful girl Friday.

As a mere spectator, I couldn't imagine the professor's personal bereavement. He had lost so much in the past few days: his two prize pupils, his precious laboratory, his hard-earned schoolhouse, his living abode, and his constant companion. Despite the tears in his eyes, he bore his suffering like the Englishman he was;

like so many brave Englishmen and women who had lost so much during the war.

The emotional moment stretched into minutes. I stood stolidly and watched the mammoth fire, mesmerized by the spectacle as well as by the potential consequences if every last spore was not destroyed. I felt a warm hand clutch mine. I glanced aside at Arlene's tear-strewn face. I didn't realize that I had been holding my breath until I felt the urgent need for oxygen. My chest heaved as I inhaled air – and hopefully air without airborne spores. Time would tell.

Suddenly I was cold, and started shivering. As I hunched my shoulders and drew my arms against my sides reflexively, I realized that it had been raining pitchforks and hoe handles (as my grandmother used to say) during the entire time that we had been outside – from the very moment when we had inspected the spores under the kitchen window. I was so distracted by dire events that my awareness of my surroundings and the state of the weather had been absent from my mind. I was sopping wet.

I forced my mental processes to expand. I tried to look past the still-burning house, across the desert, into the pitch-black sky. I shook my head, but I was unable to completely free myself of my intellectual fog. The world around me seemed to be an abstract design with little or no relation to reality. The dreadful thoughts that ran through my mind were more real to me than the physical world.

It took the raucous squeal of sand-coated brake pads to yank me out of my reverie. The Land Rover stopped in the mud next to the fire engine. Colonel Ampers and Sergeant Gompton emerged from the vehicle as if the rain were nonexistent, reminding me of a phrase from my college ROTC class: "It never rains *in* the Army. It only rains *on* the Army." That self-indulgent polemic may have been the only worthwhile insight that I obtained from the Reserve Officers' Training Corps.

"Trouble?" asked the colonel, over the pattering of

rain off his and his sergeant's ponchos.

The professor barely nodded. "Spores."

"I anticipated such. Not to worry, though. Help is on the waiy."

The radio sputtered as if on command.

"Excuse me." Colonel Ampers returned to the Land Rover and scooped up the microphone through the open window. He pulled the coil cord to its full length, then stood in the downpour as he keyed the mike.

"We got a break in the static," explained Sergeant Gompton. "We now have full mobilization. A flight of bombers and cargo planes will be here shortly to drop their loads for total eradication of the area."

The radio squawked in response to the colonel's test clicking of the transmission key. He spoke loud enough that we could all hear his end of the communication. "Holy Water. Holy Water. This is Judas Priest. Come in please. What is your position? Over."

After a burst of static, a deep-throated voice came in loud and clear. I could not distinguish the words of the responder.

The colonel referred to a map on the seat. "Holy Water. Holy Water. This is Judas Priest. Your coordinates plaice you about fifty miles southwest of the target. Proceed on your present course and home in on my open transmission. Staiy above the clouds. I repeat, staiy *above* the clouds when you drop your load. Is that understood? Over."

This time there was acknowledgment without any static.

Colonel Ampers said, "That is correct. Out." He locked open the key so the radio would transmit a continuous signal. To us, "No need to use flares to mark the spot. Not because of the burning house, but because they won't be able to see through the dense cloud layer."

I was nearly as scared now as I was when I saw the spores. "Shouldn't we be getting as far away as possible from ground zero?"

"It's too late now. Those planes are going to saturate

the area in less than ten minutes."

I gulped.

"Don't look so glum," the sergeant said grimly. "It ain't the end of the world."

I was too stunned to make a witty rejoinder. I squeezed Arlene's hand but was afraid to look her in the eyes.

"We must nip this contagion in the bud," intoned the colonel.

Or in the spore, I quipped to myself. Had I already used that line? I was too distracted to remember.

It wasn't long before I heard the drone of jet engines overhead. They were flying high and were hidden from view by the storm clouds. I didn't mind the rain. I even stopped shivering. The moment of ultimate truth was close at hand. Then I heard the Dopplering whine of a descending bomb. Only one. But one would be enough. I wondered how many atomic bombs were in Australia's nuclear arsenal, and how many they expected to use to eradicate the Mold.

I froze with resignation. A flash of light dropped through the clouds with the brilliance of a magnesium flare. The bomb struck the ground at the farther end of the orchard and buried itself in the soft loam. The light continued to burn.

Colonel Ampers grabbed the mike. "Hold Water. Holy Water. This is Judas Priest. Sighting bomb right on target. Adjust sights southward one hundred yards and commence saturation bombing. Out."

The colonel smiled. "It won't be Moldy around these parts much longer."

I kept my misplaced humor to myself: *Eliminating Mold with atomic bombs was worse than burning down a building to destroy a wasp nest.*

Then I heard the drone of many planes. Very many planes. I couldn't help but think about the numerous bombing runs that the Allies made against Germany and Japan – and those were only conventional bombs, or incendiary bombs. This was going to be worse. Much worse. Mankind was about to escalate warfare with a quantum jump: atomic saturation bombing.

The drone of the planes grew louder, achieved maximum volume, then receded into the distance. I didn't hear any bombs. I stared up at the sky, lost in thought about whether I would see the flash or the mushroom cloud a microsecond before I was obliterated. *Perish the thought*, I quipped to myself, *and the brain and the mind that thought it.*

I could no longer hear the planes. But neither did I hear the whine of falling bombs. I was befuddled. I looked to the colonel for an explanation. He held out his open palm to capture some rain. He brought his hand to his mouth, stuck out his tongue, and dipped it into the water. He smiled and nodded to the sergeant. The sergeant nodded back, as if they shared some monstrous private jape.

I couldn't help but try the experiment myself. I tasted the water and smacked my lips. "It's – it's – it's salty. It's salty! The rainwater is *salty*!"

All of a sudden everyone was tasting the water that fell from the sky like manna from heaven – *better* than manna from heaven, for it meant not just enduring life for all mankind, but enduring life for the entire world. And just as suddenly I realized what the planes carried in their bomb bays and cargo holds: not nuclear devices but tons and tons of salt. They had seeded the rain clouds with sodium chloride!

I held newfound respect for the Australian colonel and his dour sergeant. They had a sense of humor after all.

Chapter 27

Reinforcements arrived during the worst of the storm: several hundred soldiers and a fleet of vehicles from Woomera. True to the U.S. Army dictum, it didn't rain in the Australian army either; it only rain *on* the army. The officers and enlisted men ignored the downpour of saline solution, and proceeded to establish battalion headquarters in the proximity of the Aborigine village.

Colonel Ampers was in charge of the mopping up operation. He did not work by halves. Salt was being scrounged from every grocery store on the subcontinent, and more was being imported from overseas. A constant stream of trucks, either alone or in convoys, passed through Alice Springs and took the dirt road to the east.

The bridge across Noname Creek was reinforced to handle the burden of additional traffic and armored vehicles.

Paratroopers were descending from the sky like a horde of locusts, rivaling the numbers involved in the D-Day invasion. Giant cargo planes dropped food and supplies. There was even talk of building an emergency landing strip.

The colonel also took my previous suggestion to heart: he ordered napalm and flamethrowers to eradicate pockets of Mold that had escaped the deluge that had long since soaked into the soil. Every crack, crevice, and rock outcrop was double treated with fire and water, or water and fire, depending upon which demolition squad got there first.

So many tents sprung up around the village that the area looked like a refugee camp. Civilian volunteers were now supernumeraries. The best thing that we could do was to stay out of the way. The mighty Aussie military machine had the situation under control, and left us civilians to our own devices without restrictions.

Constable Harris and the civilian volunteers were given a large command tent in which to retreat from the weather, which was already beginning to moderate. Colonel Ampers asked them not to return to Alice Springs because the flow of military vehicles in the opposite direction might cause obstructions or create a traffic jam. Being Australian, the volunteers were more than willing to do whatever was necessary to aid and abet the army to expedite the massive military mission. The volunteers were well fed.

We were given our own command tent. By "we" I mean Professor Wayward, Albert and Edward, Charles and Rosalie, Arlene and myself.

"Has it occurred to anyone that all the casualties that resulted from the Mold were Aborigines?"

"Right on," Charles commented.

"An astute observation, Mr. Baker." The professor pondered my pronouncement for a moment before continuing. "It seems that the Caucasian invasion has never done any good for the local people. Aborigines who were not killed accidentally by imported diseases, were massacred intentionally by white settlers who wanted to occupy their land. The Aboriginal population has been decimated by the encroachment of the British in their bid for colonial occupation."

Charles did a masterful job of hiding his disgust. "Mikes Hitler seem like not such a bad egg after all."

Rosalie cast Charles a knowing look but did not deign to comment.

Albert Enders was pensive. "Can't say as Americans were much to blame. We moved here to mix with the locals, not to kill or conquer them."

"We figgered that marryin' was the best way to mix," Edward added.

Arlene was sullen but up to the task. "Well, I came here to write an article about the positive effect that the professor was having on the Aborigines, and not only have all my notes been destroyed by fire, but whatever I was going to write has been overshadowed by current events."

"It's just as bad for me. I came here to write a science article. Now I've got the greatest science story since Noachian flood, and I'm not allowed to write about. What good does it do to be the lucky reporter on the spot, when I can't report? And I was so overwhelmed by ongoing events that I never even got any pictures of the Mold. Boy, did I bungle this assignment. My editor will probably fire me and leave me stranded down under. He didn't furnish me with return transportation, and the plane fare exceeds my credit limit."

"My uncle will pay for your airfare home."

"Your uncle? Who's your uncle?"

"Uncle John. John Crawford. The editor-in-chief of the *Monday Morning Post.*"

Slowly I was getting the picture. "So that's how you got the job that I applied for. And that's how you got this assignment."

"Well – "

"Stop saying well and say what you mean."

"Well, uh, that is, I did major in journalism in college. After all, our family has been in the publishing business for three generations."

Something didn't ring true. "Wait a minute. The byline of the person who got the job of science reporter was Crawford, not Hawkins."

"Yes, well – I mean – I used – I did more of the research than the actual writing. Uncle John roughed out the pieces to get me started, then I rewrote them for publication. We used a pen name as a byline."

That explained the mediocrity of the pieces, but I didn't mention that. Just because a person was not a good science writer didn't mean that she wasn't a good person. Good writing was an ability that was either inborn or acquired; it bore no relation to other facets of a person's character.

I didn't let her completely off the hook, however. "Isn't is ironic that when we started this assignment you accused me of trying to take away your job, when in reality you had *already* taken away *my* job?"

"Well, I – "

"The well doesn't work because we don't have electricity for the pump."

In a huff, "You know what I mean."

"I can understand what you mean as long as I don't listen to what you say."

"Stop throwing my own words back at me."

"Stop using reflexive verbs. It's not good English."

"I didn't use reflexive verbs. I used only one of them."

She got me there, even if the point was a minor one.

"Just because our plumbing is different doesn't mean that you're better than I am, Mr. Baker."

"I didn't say I was, Miss Hawkins. I merely meant that just because you're related to an editor, and come from a long line of rich publishing magnates, doesn't automagically make you a star reporter."

"Now you're making up words, Mr. Baker."

"It's a nonce word, Miss Hawkins."

Our voices were getting louder.

Professor Wayward cleared his throat in order to break the feedback loop.

Charles chuckled. "Right on."

"There goes yer plane ticket," Edward observed. "Yer welcome to stay on our homestead."

Albert snorted. "I might be addlebrained, but I do declare that you two argue enough to be man and wife."

"That'll be the day." "Not in this century." "Marriage is for birds and birdbrains." "I'd rather marry a dingo." "And have pups the size of peanuts." "Dingoes aren't marsupial, you dingbat." "Who's a dingbat?" "You're a dingbat" "What *is* a dingbat?" "A nincompoop." "Who's a nincompoop?" "You're a nincompoop." "And you're a . . ."

There was more, but I don't remember what was said or who said it. It's just as well.

Chapter 28

After the passing of the storm, the weather in Alice Springs became warm and balmy. The hot sun dried the ground in a matter of hours. Magpies fluttered through the treetops and scrounged on the ground for crumbs and scraps – much like pigeons in New York City but not in such abundance that they were a nuisance.

Idly I wondered if Old Skinflint would be proud of me or would want to skin me alive. I was saving money on the expense account by sharing a room with Arlene. Then again, I was sleeping with the competition.

The town was now a staging area for cleanup operations. Every other vehicle was an army truck, and four out of every five people on the streets were military personnel. The fledgling army base to the east of town was growing by leaps and bounds, although it was uncertain yet that this location would become the permanent site of the proposed aerospace tracking station.

Colonel Ampers once against demonstrated that he was not without wit. He codenamed ground zero the Vatican Rag, after a satirical song of the same name that he heard on *That Was the Week that Was* – a weekly show that was televised by the British Broadcasting Corporation. The song was written by that master songster of satire, Tom Lehrer.

In addition to the massive influx of army officers and noncoms, Ampers called in a squadron of the Royal Australian Air Force that was equipped with a small fleet of brand new general purpose helicopters that were manufactured in the United States by the Bell Helicopter Company. I got to ride in one when Professor Wayward, Arlene, and I were evacuated to Alice Springs. Charles and Rosalie, and the Enders brothers, stayed behind so they could drive their vehicles to town with a westbound convoy.

The chopper pilot was so ecstatic about his new

machine that he filled me in on its history and construction details. The UH-1 Iroquois was commonly called a Huey because its original designation was HU-1 (short for helicopter, utility). It was the latest rage in the U.S. Army because its speed, range, and maneuverability made it ideal for small unit action, ground support, and medical evacuation. The cabin, or fuselage, could be fitted with seats, stretchers, and outward facing machine guns.

The pilot was assigned to the RAAF No. 9 Squadron. He was hoping to see combat if his unit was deployed to a Southeast Asian country that I had never heard of: a place where the domino theory of spreading Communism was then being felt the most. When he described the location with respect to surrounding countries, I could picture it from my social studies classes. It used to be called French Indo-China, but now it was known as Vietnam.

The Hueys were used to round up fleeing or nomadic Aborigines, and to relocate them – again – to someplace safe from Mold, falling rockets, and atomic bombs.

The Wallaby Roost seemed palatial after sleeping in a pup tent. I sat on the miniscule verandah and sipped from a mug of used crankcase oil that the Australians called coffee. The only way that I could make the brew palatable was to add two cups of hot water, half a cup of milk, and three teaspoons of sugar. This made it taste nearly as good as 10W40 before an oil change. Maybe 10W30 if the viscosity was reduced enough.

I was lost in thought about recent events when Professor Wayward emerged from the lobby and sat down next to me at the table. He had a cup of tea.

"How goes it, Professor?"

The professor looked weary – too weary to initiate conversation.

I tried to get him out of his funk. "I'd like to thank for your advice."

"Oh. What advice was that?"

"About women in general and Arlene in particular.

She still gets into a snit now and then, but once she's back to her normal self – or what passes for normal in consideration of her nature – she's almost endurable."

The professor nodded absently. "My girl was the same way."

"Your girl!" I was stunned with surprise. "I didn't know that you ever had a girl. Where is she now?"

"She is deceased."

"Oh. I'm sorry."

"Do not be. I do not suppose I deserved her anyway, after what I did to her."

I couldn't see Professor Wayward "doing" anything bad to anyone, much less his girlfriend.

"You see, she was much like your Miss Hawkins: intelligent, impetuous, perhaps even a little arrogant and snobbish – "

"Don't let Arlene hear you say that."

" – She and I vied for scholastic honors in university. Although I did not care as much for the honor as she did, I could not remain passive to her. Instead of trying to make her realize how foolish it was to compete, I had to fight her. As you have learned with Miss Hawkins, or will learn in the near future, I loved her deeply. But nevertheless I felt a compulsion to best her, to put her in her place.

"Today I realize that my attitude was primitive, atavistic – felt only by people who are immature or ignorant of life. It is difficult to realize or accept that when you are young and aggressive. By the time I understood my base motivation, she had been killed in the blitz.

"I never recuperated from the blow. I still think of her at night. Some nights. I found that I was so deeply in love with her that I could no longer form a mental picture of her face. Her features seem to have gone blank. It is true that love blinds the mind." He paused for a moment of introspection. "I live on, looking at her picture to remind me that I discovered the ultimate truth of love, but discovered it too late."

For several minutes we sat in silence. Our silence. The world around us was now a raucous noise of diesel engines, grinding transmissions, squealing brakes, and shouting soldiers.

Tears welled in the professor's eyes. I almost cried in empathy. Suddenly I felt a pang in my chest for Arlene. Desperately I tried to conjure her image in my mind, but found that the professor was right: I couldn't form a clear mental picture of her face. Love was blind indeed. And with that knowledge came the realization that I had grown fond of her despite our differences and her capriciousness. I felt that life without her might be almost insufferable. And I couldn't even remember Julie's last name.

Both of us were in a funk when Colonel Ampers arrived. I never thought that his presence could cheer me up, but it did.

"Don't get up, gentlemen." The colonel pulled a chair from under an adjacent table, swung it around backward, straddled the seat, and sat facing us across the upright back, on which he rested his elbows. When Professor Wayward indicated his cup, Colonel Ampers said, "I've already had breakfast – four hours ago."

I had a rash of comments about early birds and roosters, but kept them to myself.

"I've had a round of discussions with my superiors," he started. I almost expected to see him point a finger at the roof of the verandah, in imitation of Ronald Gadfly and his higher-ups, but he didn't. "They accepted my recommendaition to go public."

I wasn't aware that I was slouching, but his unexpected announcement made me sit up straight. "What do you mean by that?"

The colonel pursed his lips. "The situaition is too important to keep under wraps. Humanity stands at the threshold of a new campaign. We've sent rockets into spaice, orbited the Earth, and your erstwhile President maide a promise to send a man to the Moon within five years hence. As we've learned to our regret, the decontamination procedures for returning spaicecraft are woefully inadequate.

"This incident must be made a turning point in the

exploraition of spaice. I believe that we need to raise the public's awaireness of the daingers of spaiceflight con- taminaition, in order to implement saifeguards that will ensure mankind's survival against alien infections. We could try to do this by going through proper channels, but there are too many political enemas – " (Yes, he did say enemas, not enemies.) " – that would object on grounds of economy.

"We cannot afford to taike a lackadaisical attitude in this regard. What we have learned we must shaire with the world – even with the Soviets. If an incident like this occurs in Siberia, contaigion could become pandemic long before Soviet scientists are awaire that contaigion exists. Mankind could be wiped out practi- cally overnight.

"Warren, I would like you to head a newly organized task force that will be dedicated to the study of infec- tions from outer space. The government will furnish laboratories, support facilities, and personnel – right here in Alice Springs. We want Austrailia to be the leader in the study of infection from outer space."

The professor was astonished to say the least. His mouth worked soundlessly for several seconds before he cleared his throat to speak. "I – I – I am honored." Then, after a moment's thought, "A – an exobiology lab. The first one in the world. Or perhaps exobiophysics would be more precise."

"That's too cumbersome. Call it the X Lab for short."

Colonel Ampers said, "I don't care what you call it. Will you lead it?"

The professor hesitated for perhaps half a second. "Yes. Yes, I believe I will."

The colonel grinned from ear to ear. "And you, Mr. Baiker, may write about it."

"*What?*" I expostulated.

"And Miss Hawkins. We will issue a press release, of course, but I think that publication in American magazines will achieve greater circulaition and create more widespread public awaireness of the perils of pro- ceeding blindly into outer space. Especially with a

Moon landing in the works. As leaders in the space pro-
gram, your countrymen need to know this information
more than anyone. Furthermore, your articles will not
be censored nor require official approval. We want you
to tell the full story – as a warning. A very dire warning.
And not just in one article, but in a series."

I was so ecstatic that I could have kissed Colonel
Ampers. I refrained from doing so. I stuck out my hand
instead. "Deal."

His handshake was so firm that it took several minutes to realign my dislocated fingers. I might have to
dictate my first article.

Chapter 29

Later that afternoon I wandered back to the verandah of the Wallaby Roost. I felt good about the first in a series of articles about the Mold. It was more in the way of a teaser than a full-fledged piece, with the promise of more to come in the following weeks.

I had forsworn Australian coffee. I followed Professor Wayward's example and switched to tea. I pondered my future as I sipped from the mug. I saw numerous threads that were inextricably woven into a tapestry of frightening complexity and proportions. For better or for worse, my life was getting more complicated.

Arlene appeared abruptly and shattered my moment of introspection. She had turned in her bush outfit for more feminine garments that she purchased locally with money from her unlimited expense account. She wore a simple carmine print dress and matching pumps that complimented her strawberry hair and accentuated her pulchritude.

"I just got off the phone with Uncle John."

I was flabbergasted. "You called New York? Long distance?"

"Is there any other way?"

I ignored my faux pas. Ronald Gadfly would never allow such extravagance. "What did you do that for?"

"Business and pleasure. After all, he *is* my uncle. He has strong avuncular concerns about his niece in the outback in the middle of an outbreak."

"Did he know about the press release?"

"Everybody knows about the press release. It's worldwide news. That's why he left messages all over Alice Springs for me to call. Constable Harris relayed the message."

Gadfly should be so concerned about my health. "So what did he have to say?"

"He did more listening than talking. I dictated my article to the typesetter. This might be the first time in newspaper history that a weekly rag scoops the dailies."

"*You what?*" I was suddenly burning with anger. Both our magazines were published on Monday, and I had yet to turn in my article. "I thought we had an agreement – "

"Hear me out before you have a conniption, dear."

It was too late. I was already having a conniption. My face was flushed with a mixture of emotions that ranged from – I don't know what.

Arlene continued in a matter-of-fact voice that was smooth and even, as if nothing untoward had transpired. "Do you remember talking about something fishy going on at our offices? And the coincidence of both of us being given the same assignment at the same time?"

I remembered, but I was too hot under the collar to reply in other than four letter words. I kept them to myself.

"Well, it seems as if something fishy *was* going on. There was definite collusion between the managerial staff of my magazine and yours. All of it above the heads of our editors-in-chief."

I was interested, but my collar was now in flames.

"Well, Uncle John just gave me the lowdown. Our magazines were negotiating a merger."

It took a moment for me to comprehend exactly what she was saying. Should I listen to what she said or understand what she meant? I was confused.

"The merger was completed while we were having a wild time in the wilds."

My anger died down, but I was too stunned to speak. The ramifications . . .

"The new magazine entity will share subscribers and advertisers. That will increase circulation and consolidate income. Printing costs will be halved, which will help to eliminate long-term debt. Naturally the merger will result in some shifting of positions, but no layoffs are intended."

I read her last sentence as a sentence. She had an "in" with the newly organized company, while I might find myself without a job. My situation looked bleak. I calculated my chances of keeping my position as slim to none, and Slim left town. I managed to croak, "Which magazine are they going to keep? I mean, which name? The *New York Script* or the *Monday Morning Post*?"

"They're amalgamating the names. From now on we're both working for the *New York Monday Morning Postscript*."

I breathed a deep sigh of relief when she said that I was still working.

"But . . . "

"But?"

"As I mentioned, there will be some shifting of positions."

I cringed. I had a creepy feeling in the pit of my stomach. "Well?"

"Yes, I am quite well. I've just been promoted to senior science reporter."

My world suddenly collapsed around my ears. A pregnant pause extended to infinity as I contemplated the implication.

A long time later, she added, "And you've been promoted to chief editor. That makes you my boss. As your subordinate, I took the liberty of dictating your article to the typesetter, and made sure that he spelled your byline correctly."

Someday I'm going to spank that girl.

Someday. But not today.

Chapter 30

So that's the story of the Mold, the thing that challenged the supremacy of mankind.

Challenged – and lost.

There is an old adage that good will eventually win out over evil. I don't know if it's true, but that's what they say.

What they don't say, and what is implied in the adage, is that mankind represents all that is good in relation to all that is evil in the world – and off the world. It seems like a stretch to boast that mankind is inherently good when he creates such strife through war, religion, and politics.

Sometimes I wonder if mankind even deserves to vanquish what he considers to be evil, especially in light of the fact that he defined the terms.

Philosophical assertions aside, what worries me more is another manmade adage, to the effect that all scientific discoveries are eventually rediscovered. I fervently hope that this is not true, for we may not live through *this* discovery a second time.

In the back of my mind I also wonder – was the Mold truly intelligent?

Epilogue

The Mold retreated from the onslaught of destructive elements. It could not fight fire and salt-laden water, so it sought to hide.

It extended tendrils deep into the earth and far into rocky crevices. It produced buds. It launched spores into the atmosphere. It retreated from burn zones and saline drips. It surrounded its main mass with sacrificial strands that might shield the inner core.

A colloidal substance flowed through the Mold's interconnected strands. At the end of every strand existed a node. External stimuli excited each terminal node, which then triggered impulses that were transported through the colloidal substance by means of chemical reactions. As the bulk of the Mold expanded, and more strands were extended into the surrounding environment, the Mold increased in complexity. The number of nodal excitations and subsequent interactions were multiplied geometrically. Environmental information was processed internally, enabling the Mold not only to quantify every stimulus, but to respond to it as the situation demanded.

But the forces against it were mighty. The Mold lost out at every turn, was defeated at every trick. It was burned, dissolved, and corrupted.

Then a lone strand found a secure hiding place in the top of a hollow within a metamorphic boulder, where fire could not reach and where water could not flow upward against the law of gravity.

Disruptive saline solution flowed outward from the main mass and through its constituent strands, like sparks propagating along a trail of gunpowder.

The Mold signaled the separation of the strand that occupied the top of the hollow. The main mass was destroyed, but the disattached strand was saved from dissolution. It clung to the top of the hollow with instructions to wait for the water to evaporate, leaving the sodi-

um chloride behind in crystalline form - a form that was harmless because its ionic potential was zero.

There was dirt along the walls of the hollow. Water could not flow uphill, but it could be absorbed and diffused by means of osmosis, the way water was transported to the top of a tree, or soaked up by a paper towel.

The strand sensed the approach of a salt-filled droplet. It wriggled away in desperation until there was no place else to go.

The diminishing mobile droplet touched the node. The ionized solution impregnated the strand and attacked the Mold's free radicals. The essence that was Mold was destroyed. The tendril died a hasty death.

It was the last strand.

Author's Afterword

The Mold has a curious history.

I commenced my writing career in the eleventh grade, at the age of fifteen. That was when my high school English teacher offered extra credit to any student who wrote an essay, poem, or short story.

(Note: "grade" has two meanings in this afterword – the year of schooling and the level of accomplishment.)

In those days grades (or scores) were given as letters: A, B, C, D, E, and F. A was equivalent to 90% or higher; B was 80% to 89%; C was 70% to 79%; D was 60% to 69%; E was 50% to 59%; F was 49% or lower.

Each letter could also be designated with a plus or minus. A+ was the highest grade that a student could earn on a test or report card. The next lower grade was A, then A-, then B+, then B, then B-, and so on down to F-. C was considered average. E was a failing grade. F was below failing (whatever that meant).

Learning was not easy for me because I was not blessed with great memorization skill, so I studied hard and always struggled to increase my grades. The teacher's incentive to promote a student's grade by one whole letter inspired me to write a short story. Because I was already a voracious reader of science fiction, I wrote a science fiction story which I entitled "The Nothing" (reprinted in *A Different Continuum*). The length of "The Nothing" was 47 double-spaced typewritten pages, or some 16,000 words.

The teacher was so impressed by my work and its length that he raised my grade from B to A.

Other things were different in those days: (1) grade years were split into A and B; one had to achieve a passing grade in 11A before being promoted to 11B. If you *didn't* pass 11A, you didn't get promoted to 11B. (2) The grades started twice a year: September and January. This is much like college semesters except that A and B also started twice a year. Thus you could start A in January, not only in September. This also meant that if you flunked a grade, you only had to repeat half a year of school, not a whole year. (3) If a student's birthday fell before the end of the next semester, he could start first grade nearly half a year early.

What this all boils down to is that I started 1A in January at the age of just over five and a half, because my birthday fell on June 2, and school ended around June 21 or 22 (depending upon the number of snow days). This means that throughout my school years I was younger (and less mature) than my classmates. It also means that I graduated high school in January, then had to idle my time for the next seven months because colleges started only in September.

After writing "The Nothing" in 11A, I wrote several short stories during summer vacation and during 11B and 12A. I was still sixteen when I started my first novel, *A Journey to the Center of the Earth*. I completed the novel during summer vacation between 12A and 12B, by which time I had turned seventeen.

In the autumn of 1963, during 12B, I conceptualized *The Mold*. After graduating from high school in January 1964, while I was still seventeen years old, I wrote the initial drafts in a somewhat haphazard manner. First I wrote 43 pages in long hand to about the middle of the story. Then I stopped and began a second draft on a manual typewriter (after attending a two-week typewriting class at a local business school, in preparation for college). This time I wrote (or typed) 74 pages, continuing past my previous stopping point so that the latter part was a first draft. I stopped again before completing the story, backtracked to typewritten page 45, and wrote a second draft of pages 45 to 74, then completed the story in first draft. This new section I numbered a1 to a55.

The total number of typewritten pages that comprised the completed novel was 100. The final word count was approximately 43,000 words. It was nearly summer by the time I finished *The Mold*. I went to work as an electrician for my father (who was an electrical contractor) until classes commenced at Temple University.

In 1966, I was drafted between semesters and sent to Vietnam as an infantryman. After my discharge in 1968, I joined the electrical union, took a four-year apprenticeship course, and worked as a commercial and industrial electrician. Construction was slow in 1976. I was unemployed for much of that year and the following two years. I used my off-time to start writing science fiction again.

In 1979, I retired from the union and started working fulltime as an author: an occupation that I have continued to the present day.

The Mold has never been far from my conscious thoughts. I had often thought about writing a final draft for publication. In 1982, 18 years after I wrote the first two drafts, the memory of *The Mold* suddenly erupted into the forefront of my mind. I was inspired to write a new beginning that would blend in with the original story. Without rereading the original manuscript, I wrote two chapters totaling about 12,000 words. Then I stopped just as abruptly as I began.

This kind of monster novel was by then out of vogue. I had not lost interest in the book; I simply recognized that I probably could not find a publisher for it. So I put it aside. I filed the new manuscript with the original version, which together totaled some 55,000 words. There it lay fallow until 2009. Then I brought out all my drafts and reread them.

I found some interesting peculiarities in addition to the odd arrangement of the 1964 drafts, which stopped and restarted so inconsistently. Once I ironed out the storyline, and rearranged the text so that it flowed from beginning to end, I discovered that my memory had been somewhat remiss. In the new but abortive beginning, I had remembered the characters but not their names. I had created new names for them. In 2009, I consolidated the material and reverted the names to those in the 1964 drafts.

Instead of simply reworking and re-editing the original and subsequent versions, I commenced to rewrite the entire book. I used the old manuscripts as guidelines, but I added new material and invented new characters, situations, and scientific analyses. The result is a book that is now some 78,000 words in length, and that is more fulfilling than the original, because I now had the benefits of history and hindsight.

For example, as a teenager I knew that Woomera was a rocket testing site, but I had no idea that the Australians detonated atomic bombs there. That was probably a closely guarded secret; or, at the very least, the information was not readily available in the United States.

I did a considerable amount of research in preparation for this rewrite, in order to add greater verisimilitude to the story. What was secret in 1964 is now public information. After learning the truth about the nuclear bomb tests, I was able to work those facts into the storyline. This added emphasis and fictional meaning to the purpose of the real-life nuclear detonations.

At this remove, I have also had the benefit of visiting Australia and backpacking in the wilderness. In 2001, I was invited to speak at a technical diving conference in Melbourne. In 2002, I was invited to speak at another technical diving conference in Sydney. In each instance I extended my stay so I could dive on shipwrecks and take in some of the local flavor. Thus I saw the Australian outback firsthand.

The most difficult part about not updating the timeframe to the present was in avoiding anachronism. I had to be careful not to allude to events that had not yet occurred in 1964, and not to mention technology that did not exist at that time. Additionally, some facts that I knew at the time I have long since forgotten. I had to research current events in 1964 so my characters gave the appearance of familiarity with their world the way it was.

I have never adopted a specific writing style. Rather, I let the tone of each book establish an individual style for that particular book. This stylistic eclecticism means that some books are formal, some are informal, some use contractions, some avoid contractions like the plague (except in dialogue), and so on.

Ironically, much of the juvenility with which Tim Baker is imbued, I added in this final draft. The original story was written informally but not in so much of a conversational or wise-cracking style. Despite the elements of horror, I wanted to make this a lighthearted story that delved into the personalities of the protagonists and their complex relationships with some fun. I did this partially by adding some boyishness to Tim Baker's character. Also, by writing the book in the first person singular (which I did originally), I could get away with awkward sentence structures, because the story reads the way in which Tim Baker would have told it verbally. Thus there are sentences that end with prepositions (a definite no-no in formal writing), and other unconventional grammatical constructions. This lack of formality in language, sentence structure, and grammatical construction was intentional.

Throughout my writing career, all my imaginative concepts have come strictly from inspiration, much like a flash of insight. I have never consciously tried to create an imaginative concept. *The Mold* is no exception. It erupted spontaneously. As I wrote the final draft, many new concepts popped into mind, making a finished product that is greater than the summation of its three time-spread parts.

From beginning to end, *The Mold* was 45 years in the writing. Not many authors can claim (or are willing to admit) that it took them nearly half a century to write a book. In my defense, however, I can call attention to the fact that I wrote 50 books in between.